Nation and Region in Modern American and European Fiction

Comparative Cultural Studies,
Steven Tötösy de Zepetnek, Series Editor

The Purdue University Press monograph series of Books in Comparative Cultural Studies publishes single-authored and thematic collected volumes of new scholarship. Manuscripts are invited for publication in the series in fields of the study of culture, literature, the arts, media studies, communication studies, the history of ideas, and related disciplines of the humanities and social sciences to the series editor via email at <clcweb@purdue.edu>. Comparative cultural studies is a contextual approach in the study of culture in a global and intercultural context and works with a plurality of methods and approaches; the theoretical and methodological framework of comparative cultural studies is built on tenets borrowed from the disciplines of cultural studies and comparative literature and from a range of thought including literary and culture theory, (radical) constructivism, communication theories, and systems theories; in comparative cultural studies, the focus is on theory and method as well as application. For a detailed description of the aims and scope of the series, including the style guide of the series, link to <http://docs.lib.purdue.edu/clcweblibrary/seriespurdueccs>. Manuscripts submitted to the series are peer reviewed followed by the usual standards of editing, copy editing, marketing, and distribution. The series is affiliated with *CLCWeb: Comparative Literature and Culture* (ISSN 1481-4374), the peer-reviewed, full-text, and open-access quarterly published by Purdue University Press at <http://docs.lib.purdue.edu/clcweb>.

Volumes in the Purdue series of Books in Comparative Cultural Studies
<http://www.thepress.purdue.edu/comparativeculturalstudies.html>

Thomas O. Beebee, *Nation and Region in Modern American and European Fiction*
Paolo Bartoloni, *On the Cultures of Exile, Translation, and Writing*
Justyna Sempruch, *Fantasies of Gender and the Witch in Feminist Theory and Literature*
Kimberly Chabot Davis, *Postmodern Texts and Emotional Audiences*
Philippe Codde, *The Jewish American Novel*
Deborah Streifford Reisinger, *Crime and Media in Contemporary France*
Imre Kertész and Holocaust Literature, Ed. Louise O. Vasvári and Steven Tötösy de Zepetnek
Camilla Fojas, *Cosmopolitanism in the Americas*
Comparative Cultural Studies and Michael Ondaatje's Writing, Ed. Steven Tötösy de Zepetnek
Jin Feng, *The New Woman in Early Twentieth-Century Chinese Fiction*
Comparative Cultural Studies and Latin America, Ed. Sophia A. McClennen and Earl E. Fitz
Sophia A. McClennen, *The Dialectics of Exile*
Comparative Literature and Comparative Cultural Studies, Ed. Steven Tötösy de Zepetnek
Comparative Central European Culture, Ed. Steven Tötösy de Zepetnek

Thomas O. Beebee

Nation and Region in Modern American and European Fiction

Purdue University Press
West Lafayette, Indiana

Printed in the United States of America.

Library of Congress Cataloging-in-Publication Data

Beebee, Thomas O.
Nation and region in modern American and European fiction / Thomas O. Beebee.
p. cm.— (Comparative cultural studies)
Includes index.
ISBN 978-1-55753-498-9
1. Setting (Literature) 2. Space and time in literature. 3. National characteristics in literature. 4. Fiction--History and criticism. I. Title.
PN3383.S42B43 2008
809.3'9358--dc22
2008013003

To Alice and José, for kindly accepting me as a new native son and making me a part of their landscape

Contents

Acknowledgements

The author wishes to extend his sincere thanks to the Institute for Arts and Humanities of the Pennsylvania State University for its support of much of the research and travel necessary to complete this book. Sharon Bailey, Manuel Fernández, Steven Hunsaker, and Angelina Ilieva provided unflagging help with research and documentation. I am especially grateful to Monika Giacoppe for first acquainting me with several of the texts treated in this study, as I am to Oleone Coelho Fontes and Wilton de Carvalho for acquainting me with the landscapes of Canudos and of the *sertão* in general. My gratitude is also due to the anonymous reader of the manuscript of this book for the many helpful suggestions, as well as to Steven Tötösy de Zepetnek for his leadership in comparative cultural studies in general and his support of this project in particular.

Several chapters of this text are based on previously published articles. A version of chapter 2 appeared under the title "Ways of Seeing Italy: Landscapes of Nation in Goethe's *Italienische Reise* and its Counter-Narratives," *Monatshefte* 94.3 (2002): 322-45. A version of chapter 4 was published under the title "Talking Maps: Region and Revolution in Juan Benet's *Volverás a Región* and Euclides da Cunha's *Os Sertões*," *Comparative Literature* 47.3 (1995): 193-214. A version of chapter 5 appeared under the title "Tryptichs of Solipsism: *The Sound and the Fury* by William Faulkner and *Fogo morto* by José Lins do Rego," *The Comparatist* (1999): 63-88. Permission to reproduce from these sources is gratefully acknowledged.

The author also gratefully acknowledges the following graphic images: Figures 1 and 2 of chapter 1 from Peter Gould and Rodney White, *Mental Maps*, 2nd ed. Boston: Allen & Unwin, 1986, pp. 60, 86. Figure 3 of chapter 1 from Richard Johnson, "What is Cultural Studies Anyway?" *Social Text* 16 (1986): 47, Copyright © 1986 by Duke University Press. All rights reserved. Used by permission of the current publisher. Figure 4 of chapter 2 was drawn by Palavina Ferreira Beebee-Galvão, and her permission to reproduce it is gratefully acknowledged. Figures 20 and 21 of chapter 7 from *Secrets from the Center of the World* by Joy Harjo and Stephen Strom. © 1989 Arizona Board of Regents. Reprinted by permission of the University of Arizona Press. The topographic map in Figure 19 is courtesy of the US Geological Survey. Unless otherwise acknowledged, other images appearing in this book are by the author.

Chapter One

Introduction

In this book I explore comparatively the role of literature in the production of national, regional, local, global, and glocal mental maps. (On the concept of the glocal, see below, 175). The time frame for investigation is, roughly, modernity, defined as having "the core feature that the world is divided into sovereign nation-states with no remainder" (During 98). Literature has had a role to play in how that division is made, perceived, communicated, and deconstructed. Literature has also questioned whether there is indeed "no remainder." In moving through the scales mentioned above, then, for our purposes the nation-state will remain the *metron*. Johann Wolfgang von Goethe's writing stands *ante portas* of the modern era, while Rolf Dieter Brinkmann's mental maps are deliberately postnational. Other writers, such as Ivan Turgenev and Euclides da Cunha, devotedly make use of the nation-state *metron* but consistently find local "remainders" that trouble their discourse. It is significant that Simon During devotes an entire chapter of his introductory work on cultural studies to the topic of "Space" (79-106). It is equally significant that his *metron* for discussing space is not the nation-state, but globalization. During notes that spatial considerations have replaced those of chronology in analyses of development, economics, and so forth. He further observes that "cultural studies' relation to space requires further thinking through" (106). Indeed. If cultural studies scholarship is to have a scholarly purpose, beyond merely the ideological one of either celebrating the supposed passing away of the nation-state or harpooning the Leviathan of globalization, then we need detailed analyses of the role played by space in the construction of national identity.

In this study, I proceed from the following methodological considerations, the terms of which will be explained in the rest of this chapter: 1) A number of literary narratives of the last two hundred years, written in a number of languages and on at least three continents, have as a goal the drawing of "mental maps" for their readers. I have narrowed my concerns to mental maps of heterotopias, fictional places that invoke readers' lived experiences of space and place; 2) As the most powerful and contested of meanings attributed to place in the modern era, and one that links space and topography with social events and relations, nation will be a key term for analyzing the operation of heterotopias. "Nation" is to be understood differentially, against other organizing concepts such as region, empire, locale, and so forth; 3) My main tool for analyzing mental maps will be Hayden White's mimesis-diegesis-diataxis triad. In particular, diataxis represents the rhetorical strategy for using spatial elements to create narrative, and vice versa; 4) The diataxis of literary mental maps reveals itself best when considered alongside other types of mental maps, and indeed alongside cartography and chorography proper. The unifying point is that all mapping strategies are cultural productions. Among the many types of maps, mental and otherwise, literary ones distinguish themselves through the rich detail they provide on private perception, their fluidity of movement from mimesis to diegesis and vice versa, and their foregrounding of diataxis. The "protagonists in motion" of these literary works, through their traversal of space, topography, and social networks, confront and (re)create their mental maps of nation and region; 5) The cultural production of mental maps involves a cyclical interchange between private perceptions and public forms. For the former, authorial biography, interviews, information on sources, and so forth provide valuable data; for the latter, reviews, counternarratives, and other receptions of the works in question are useful. For some works I provide my personal narratives of the places in which they occur. These are provided not, of course, as the more real or positivist reports on the reality of these places, but as alternative diataxes that add another comparative perspective to the analysis; and 6) Through comparative analysis across a chronological and cultural range, this study contributes to our understanding of the process of mental mapping in literature, including the impingement on it of historical forces and ideologies and its interaction with other discursive practices. I hypothesize that the well-known ("canonical") authors and works of this study derive their prominence in their respective national literatures not only from stylistic and narrative excellence, but in part as well from the imaginative power with which they engage the issues discussed above. Such works provide the optimal study examples, due also to the relative abundance of data discussed in number 4. The basic conceptual tools of comparative analysis can best be brought to bear by choosing works from a variety of languages, cultures, and periods. Observation of the *differentiae* provides salience to the above theses.

Steven Tötösy de Zepetnek has proposed a "set of [ten] principles" for the development of comparative cultural studies (255). I will not address all ten here, but I do wish to emphasize how my work adheres to the ones that in my opinion are most transformative of received notions of comparative literature. First, I place emphasis on the "how" rather than the "what" (259) of mental maps in literature, of how certain places and landscapes become narrativized. The intersection of my study with geography and history fulfills the fourth principle, "to study culture in its parts . . . and as a whole in relation to other forms of human expression and activity and in relation to other disciplines in the humanities and social sciences" (260). Finally there is the call for "evidence-based research and analysis" (260) and a "systemic and empirical approach" (261). This point echoes During's about the need for actual studies of how "cultural studies' relation to space requires further thinking through," the "thinking through" involving in this case detailed study examples that can be verified and compared with each other. In positioning my work vis-à-vis other scholarship, I should first of all explain the limits I have put on the type of mental maps to be considered: heterotopias.

Heterotopias

"Queequeg was a native of Kokovoko, an island far away to the West and South. It is not down in any map; true places never are" (Melville 852). Ishmael's dictum in *Moby-Dick* reminds us that places' most substantial truths remain invisible to mapping. Conversely, Huck Finn's exclamation, "We're right over Illinois yet. . . . Illinois is green, Indiana is pink. . . . It ain't no lie; I've seen it on the map, and it's pink" (23), shows the power of imagination—in the literal sense of mental imaging—in constructing what most would agree is an overly simplistic sense of "imagined community": one is either a pinkie or a greenie. For Twain's Huck Finn, truth is a function of mapping, whereas for Melville's Ishmael it is what inevitably escapes mapping. Melville invokes utopia, the no-place, while Huck only perceives of place as a collection of physical *accidentiae* that impinge on the senses. When one considers the irony that Kokovoko does not physically exist at all (a status it shares with its affirmer, Ishmael, who spells it Rokovoko in the next chapter), perhaps the two statements coincide. We may laugh at Huck's attitude to maps, but must take seriously people's deriving their sense of *eretz Israel* from scriptural narratives, or of Kosovo from legends of a battle that took place there centuries ago and has left no physical traces. That both reflections arise in the context of eccentric journeys—Tom and Huck are taking a balloon trip—and out of confrontations with an Other cannot be mere coincidence.

This book explores and compares several verbally constructed heterotopias in American and European literatures of the modern period. I derive the term "heterotopia" from the brief but highly suggestive essay by Michel Foucault, "Of Other Spaces." After establishing the historicity and discrete functionality of different social spaces, Foucault posits two complementary categories of sites that "have the curious property of being in relation with all the other sites, but in such a way as to suspect, neutralize, or invert the set of relations that they happen to designate, mirror, or reflect" (24). These two sites are the utopias, which exist exactly no place, and the heterotopias, that is, the "counter-sites, a kind of effectively enacted utopia in which . . . all the other real sites that can be found within the culture, are simultaneously represented, contested, and inverted" (24). Under utopias I understand imaginary places, while under heterotopias I understand what I call "true imaginary places." There is however a degree of redundancy to the term "true imaginary place," because some geographers define "place" as "a segment of space which an individual or group imbues with special meaning, value, and intention" (Parmenter 4). In other words, place is a true segment of space endowed with imaginary qualities—again using "imaginary" in the sense of imaging something rather than perceiving it, as in the act of remembering. We can only imagine, not perceive, meanings, values, and intentions. W.J.T. Mitchell, in introducing *Landscape and Power*, draws on the work of Michel de Certeau and Henri Lefebvre to reverse this distinction in his discussion of space, place, and landscape as a "dialectical triad, a conceptual structure that may be activated from several different angles. If a place is a specific location, a space is a 'practiced place,' a site activated by movements, actions, narratives, and signs, and a landscape is that site encountered as image or 'sight'" (x). *Landscape and Power* is a volume of essays that collectively analyzes the social relations that inform, promote, and demystify landscape as a genre. While the texts I analyze function occasionally as landscapes, the key object of my investigation is their narrative and rhetorical construction of mental maps of their respective heterotopias. Such constructions may be made in a number of media, including literary texts.

Foucault notes that a mirror epitomizes the concept of heterotopia, since it opens a space both absolutely unreal and at the same time constitutive of reality as placedness. A hall of mirrors is both an illusion, in that the mirrors artificially expand the physically available space, and also reality, inasmuch as the mirrors inevitably help constitute that space as it is lived and perceived. (Utopia, on the other hand, is more analogous to a painting or drawing.) We may resume the mirror's combination of fiction and reality by saying that the mirror exemplifies the mapping of space, especially because here the mapping process is taken to the extreme of fidelity: the mirror maps point by point, but distorts nonetheless. In addition, the mirror image of real space becomes a part of that space as lived and perceived.

As I have used the term in this book, heterotopias are "virtual" spaces that remain recognizable and meaningful to those who experience them. They are true imaginary places. In *Postmodernist Fiction*, Brian McHale has provided a somewhat different definition of "heterotopia," as "the disorder that is made up of fragments of a number of incommensurable orders" (163). One must experience, McHale argues, a measure of cognitive dissonance in one's apprehension of place in order for it to qualify as a heterotopia. (It seems that Foucault deliberately left incommensurate clues as to what he meant with his conception.) The two definitions could be reconciled by our agreeing that every fiction engenders cognitive disorder, as readers compare the "verbal map" of the text with their own mental maps. Just as dissonance in search of resolution provides the interest of a melody in music, so too the cognitive dissonance of heterotopia focuses readers' attention on nation, as they weigh the similarities and differences of their own spaces and communities against the fictionalized ones studied here.

For this study, I have chosen authors whose fictional worlds map onto real places in a nearly one-to-one fashion, mirroring reality, but with the increased focus and self-reflexivity that fiction provide. The Middle Earth of J.R.R. Tolkien, for example, or the locations of Sinbad's travels, I consider as fantastic geographies rather than true imaginary places. My distinction between utopias as imaginary places and heterotopias as "true imaginary places" corresponds to Neil ten Kortenaar's distinction between what he calls imaginary and fictive settings: "In the space of imaginary settings, at least some rules of scale and of nature do not apply. . . . The fictive setting presumes at once a fully mapped world *and* room to add to that map. . . . No reader of George Eliot will look for Loamshire on a map of England. Readers will trust that they already know the fictive setting's essential features; its character can be derived by extrapolating from the known series of central English counties" (230-31). Kortenaar concludes from this that for settings in Europe and North America, nation-states cannot be pseudonymous; that is, writers will not give a fictional name to a country that behaves in substantial ways like a real country. There is presumably no room to extrapolate on the level of nation the way there is on the level of locale. (Kortenaar's overall thesis is to make a North-South contrast in this regard, detailing in particular the situation in African literature, where nations tend more towards the imaginary and pseudonymous, presumably due to the much weaker performance of nation-states on that continent.) Indeed, a good way of describing the group of texts I have chosen for this study is that real place names appear almost exclusively. The exceptions, such as Juan Benet's "Región," are of course what are worth observing closely, as are the authors' politics of naming (a form of diataxis). Mary Austin and William Faulkner love to give exact place names, for example, but avoid mentioning that their narratives take place within the United States.

Phillip Wegner's study of the genre of utopia (with study examples by Thomas More, Evgeny Zamyatin, George Orwell, and Ursula K. Le Guin, among others) and of its relation to ideas of nation, which like the present work explores the mutual influence of spatial orientation and narrative, argues that utopic narratives "mediate . . . between the world that is and the world which is coming into being" (37). This orientation towards the future, rather than present or past, or the escape from categories of time altogether, does seem to me to characterize utopias in contradistinction to the heterotopias I examine, where future configurations of territory and identity are bracketed in favor of "securing the border" or finding one's way. An interesting overlap occurs, however, in the post-World War II narratives of José Maria Arguedas and Nicole Brossard, both of which foreground processes of deterritorialization and "smooth space" so as to invoke notions of u-topia in its literal sense of "no-place."

Nation as heterotopia

In the modern period, perhaps the most powerful and contested "special meaning, value, and intention" associated with space has been the idea of nation. In *The Production of Space*, Henri Lefebvre discusses nation as the intersection of two social moments—markets and violence—that "combine forces and produce a space" (112). Unfortunately, he does not develop this thesis further. My fencing off of a particular place-relation in this way joins a long list of predecessors. E. Relph, for example, distinguishes kinds of space on a continuum ranging from "primitive" to "abstract," while Yi-Fu Tuan has a correlative but differently indexed range, from "hearth" to "cosmos" (Relph 2-28; Tuan, *Cosmos*). Otherwise, division by threes is quite popular, from J.N. Entrikin's "subjective," "objective," and "cultural" spaces, to Robert Sack's places that stress either "nature," "social relations," or "meaning," to Edward Soja's "perceived," "conceived," and "thirdspace" (Entrikin, *Betweenness* 55; Sack, *Homo* 84; Soja, *Thirdspace*). Wesley Kort has concluded that modern fiction determines three types of place-relations: "cosmic or comprehensive; social or political; and personal or intimate space" (*Place and Space* 150). The verbal maps of this study are distinguished by their ability to cut across and even unite these disparate categories.

The recursive appearance of "nationalities" within the second definition (labeled "1.b") of "nation" given in *Webster's Third International Dictionary* indicates the layered historical development of the concept: "a community of people composed of one or more nationalities with its own territory and government" (1505). In the "Keywords" section of her book on exile literature, Sophia A. McClennen uses a similar definition for the more specific and historically bounded term, "nation-state": "The nation-state refers to the physical borders of a country as well as those formal institutions responsible for its existence" (26). Matthew Sparke alerts us to the significance of the hyphen linking nations with states: "As a text-spanning symbol of space-spanning phenomena, the hyphen in nation-state came to represent two mutually reinforcing geographical processes. On the one side were the diverse state practices such as border policing, migration control, and planning that regulated territorial belonging. On the other side were the modern space-producing social and cultural dynamics that, in generating taken-for-granted national landscapes, national monuments, national maps, and so on, gave state regulation its space and place of legitimacy" (xiii). Territory and boundaries, in this view, are not ontological givens, but rather cultural productions needed by the state in order to become a nation. Narrative, too, produces space in order to become, and this book examines cases where this latter process of the coming-into-being of literary fiction intersects with national projects, and in so doing, reveals something about those projects.

In this section, I explore some of the connections between the development of chorographic representations of space, and the evolution of the "nation-state" idea ("Chorography" refers to the art of delineating different countries or districts). Originally, the term "nation" referred to what today we call a clan, tribe, or ethnic group, a community into which one was born (hence the derivation from Latin "natus," the past participle of "to be born"). I consider such communities to be somewhat less imaginary than the nation-states described above: there is more probability—just do the math—of knowing (even if only by hearsay) a significant portion of other members of a clan, tribe, or ethnic group than other citizens of a nation. Indeed, the use of "community" to refer to a nation stretches the boundaries of that term, because members of a nation only meet a tiny fraction of its other members. Benedict Anderson's approach, which notes that nations are formed when people partake of the same set of symbolic landmarks, be these physical or literary, counters empirical descriptions of national identity as "collections" of customs or laws—or, as in the "anthropogeography" of Friedrich Ratzel, as the product of geographical features that mix and separate peoples. How, then, is this sense of common territory achieved and communicated? Two features of national territory make it an imaginary construct. One is the principle of "difference": territory is determined by borders, which themselves belong to none of the nations that share them. "Our" national territory is determined by what belongs to "them" or to the "Other." The second feature is the mediated perception of territory. I take as axiomatic that, other things being equal, a sense of living with a group of people in proximity within a bounded territory tends to convey a sense of belonging to that group. For example, the US school song "America the Beautiful," tells US-Americans that their nation occupies a territory extending "from sea to shining sea," a concept of expansiveness and continental dominance that closely corresponds to the phrase "From the Memel River to the Etsch" ("von Memel bis Etsch") in a part of the German national anthem no longer sung. These rivers are located in East Prussia and the Tyrolean Alps, respectively, and their naming constitutes an aggressive claim on land populated largely by non-German speakers (Memel today is called Klaipeda and the Etsch retains its Italian name of Adige). Homi Bhabha and the deconstructionists, on the other hand, would interpret national territory as a sequence of differences. The idea of nationhood is thus constructed through a series of comparisons and confrontations with other entities that lie beyond—and sometimes within—the border. Such comparisons and confrontations determine and define the true imaginary places depicted by the authors treated in this volume. The question then becomes, how is that sense conveyed to members of the group? Unlike language, for example, which co-nationals participate in daily and directly, few if any citizens travel every inch of their national territory. Fractal theory tells us that it is impossible to account for every nuance of a frontier or a shoreline, even if these were not constantly shifting as they are with the tide and winds. To gain a sense of the extent, shape, wholeness, and contiguity of a national territory, we must rely on a variety of symbolizations, including distinctive natural features and maps. Governments (especially tourism agencies) provide their nations with a variety of chorographic maps. Several canonical works of literary fiction, I argue in this book, have created for their readers chorographic mental maps that have a bearing on national history and identity, although their purpose and use is of course radically different from those of the tourism agency.

To appropriate the title of one of the seminal essays of modern political geography, the state is an important (although not the only) *Landschaftsgestalter*, or creator of landscape (Hassinger). As Edward Saïd has argued, the general phenomenon of imperial expansion pushed geographical exploration in its vanguard and trailed it in its wake; this geographical awareness, more or less articulated in various writers, appeared in cultural productions that cross-fertilized each other across disciplines. Geography in some instances became a kind of "super-Orientalism" used to distinguish "knowable" from "exotic" regions. "Structures of location and geographical reference appear," writes Saïd, "in the cultural languages of literature, history, or ethnography, sometimes allusively and sometimes carefully plotted, across several individual works that are not otherwise connected to one another or to an official ideology of 'empire'" (*Culture and Imperialism* 52). In this study I discuss nationalist and regionalist rather than imperialist senses of place. The latter has received a great deal of attention, in addition to Saïd, in the collection *Landscape and Power* mentioned above, and in work by Mary Louise Pratt and Mohammed Bamyeh. This thesis has also been worked out in studies of particular areas, for example by Ralph Bauer for the Americas, Tom Conley for early modern France, Matthew H. Edney for British India, and Ricardo Padrón for early modern Spain. The thesis common to these works involves positing an "imperialist gaze" in maps and landscapes that are produced according to an agenda of dominance and appropriation. Throughout this study, the bird's-eye or "establishing-shot" view necessary to such an agenda is contrasted with the traversal of a labyrinth that corresponds, in my view, with nation or locale, rather than with empire. The two senses of place, national and imperialist, are of course intertwined with each other, and with other dimensions such as the local and the global, as we will see especially in the chapter on Russian mental maps and on Euclides da Cunha, and as is patent in the role cartography played in the history of the "discovering," parceling out, and settling of the Americas.

Mental maps

Most of us link automatically the signifier "map" to a visual representation of geographical, political, and historical realities. Whereas the Latin *mappa* indicated the napkinlike piece of paper on which the map unfolded itself to view, the French word *plan*, which indicates both "map" and the English cognate "plan," is more indicative of what mapping has been for most of human history: a plan for getting around, for interacting socially and with the environment. Our intuitive identification of the map with lines and figures complements our reluctance to recognize maps as texts. Yet, as Paul Bohannon points out, "the Western map is . . . a strange kind of map. All peoples of the world have maps of one sort or another—usually they are not written, but the raw material is there for a 'map,' a view or image of the terrestrial world. None save modern technical civilizations have maps in which precision is so essential. There are a few peoples who divide up the world by natural boundaries such as rivers and hills. Most, however, see it in terms of social relations and the juxtaposition of social groups" (176). Social relations and juxtapositions, however, are more susceptible to narrative than to spatial description. Our relative neglect of other forms of mapping in favor of the visual and concretely spatial is intimately bound up with ideas of property and belonging, which in turn lie at the basis of the nation-state. So economically and politically powerful have such ideas become that, as Edward Soja points out, "it has become extremely difficult to accept the fact that not all the world's people share this perception of space, that in many societies clearly delimited boundaries assumed to have some permanence were, until recently, virtually unknown, and that large areas could and did remain acceptably unorganized and outside the jurisdiction of any group" (*Political* 10). To repeat During's point, the modern era divides the terrestrial world up between various nation-states, "with no remainders." Naturally, the verbal maps of literary texts lack the *mathesis* and precision of chorographic maps and they retain much of the "acceptably unorganized," as for example in the constant orientational confusions of Nikolai Gogol's *Dead Souls*, which, as I will show, is a text about the great art of getting lost. In most of the texts to be examined, the *mathesis* of correct orientation confronts or is subverted by the erratic taxis of a "solitary hero," a conflict that can be linked in each case to the ideological issues of nationality that are also in play.

Perhaps the simplest mental map is the place name: what is the mental image conjured by the utterance of a place name like "Ukraine," "Brazil," "Kokovoko," or—Juan Benet's sly coinage—"Región"? The sharing of names such as these, and of the assumption of mutual understandings about what is meant by them, provides a ready example of Anderson's "imagined community." Yet the diversity of processes needed for the construction of such a sign, when brought to the surface, make any straightforward and unambiguous reading of it impossible. Homi Bhabha states the difference thus: "The recurrent metaphor of landscape as the inscape of national identity emphasizes the quality of light, the question of social visibility, the power of the eye to naturalize the rhetoric of national affiliation and its forms of collective expression. There is, however, always the distracting presence of another temporality that disturbs the contemporaneity of the national present" ("DissemiNation" 295). Bhaba notes the uneasy coexistence of naturalized with narrativized landscapes. Many place names would lose their evocative and what Bhabha calls their "pedagogical" qualities if the etymological entries describing their histories were to gain the upper hand in their inhabitants' minds. Since, as Friedrich Nietzsche has claimed, "only something which has no history is capable of definition" (71), the narrativization of landscape can easily disrupt the pedagogical dimension of national discourse. The difference is at once analogous to the distinction I am making between chorographic maps and the mental maps of literature, and to the presence of both mimetic and diegetic strategies within them.

The verbal map is one form of what Peter Gould and Rodney White have called the "mental map." In their book of this title, Gould and White take verbal or conceptual data and represent it graphically in (for the most part contour) maps. In other words, they provide chorographic images based on mental maps. These maps bear a resemblance to humorous maps such as the famous "New Yorker's (originally, the Bostonian's) map of the United States," in which New York (or Boston) takes up 80% of the space and California borders it, with a small, uninteresting wilderness in between. That is, the map's distortions (and, as Melville and Twain both suggest, all maps are distorted, or we might say that distortion is inherent in the notion of a map) are affective: places that are loved by or known to the New Yorker appear larger and closer than if they had been outlined by a neutral party or method. Similarly, Gould and White make the desirable areas of their interviewees higher or darker than *terra incognita* or *terra non grata*. The creation of mental maps is an important part of that transformative labor that turns environment into landscape. Like the more physical

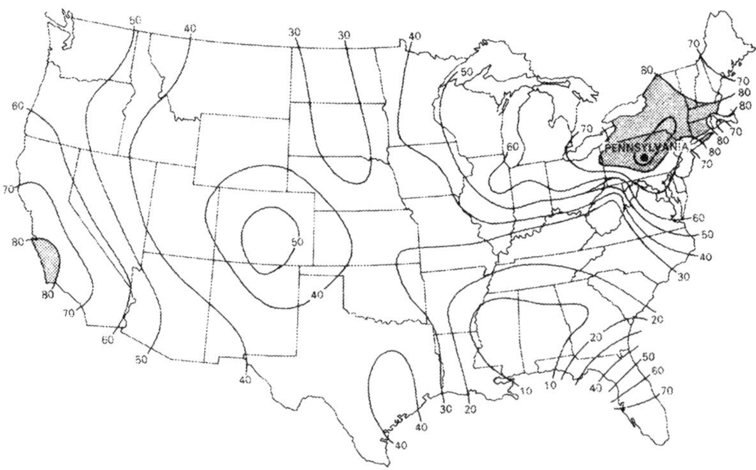

Figure 1. The Mental Map of the US from Pennsylvania (Gould and White 86)

processes of mining and lumbering, cultural mapping takes what is given, the data, and works them to the benefit of individuals and of society.

Gould and White asked people in Pennsylvania to rank places in the United States in terms of desirability. The rankings were then mathematically converted into isobars (see Figure 1). Rather than using size to indicate the affective impact of a region, as in the New Yorker's map, Gould and White contrast high with low. We see high areas, in the eighties and nineties out of a possible 100, in and around Pennsylvania itself, in California, and in Colorado—the celebrated "Rocky Mountain High." We also observe a "Southern trough," showing that states such as Mississippi and Alabama do not appeal to Pennsylvanians. Gould and White also interviewed people in Alabama, and the mental map derived from those interviews shows a certain symmetry to that of the Pennsylvanians (see Figure 2). California and Colorado continue to receive high ratings (no pun intended), and of course Alabama has substituted Pennsylvania as a very desirable place to live. The two maps show steep declines starting at the Southern and Western borders of the states where the interviewees live, respectively. Pennsylvanians think their own portion of the Appalachians is desirable, but not that pertaining to their neighbors in West Virginia. Alabamans vastly prefer their own portion of the Deep South to that occupied by Louisiana and Mississippi.

We might say that these contour lines represent in part the kinds of stories Americans tell each other about the various regions of their country, and perhaps also their memories of these regions. We see in Gould and White's presentation only the map, but not the stories, images, and other cultural information that people have absorbed in order to construct their mental maps. A mutual conditioning occurs between the narratives we construct about certain places and regions, and the overall pleasure or displeasure we experience in imagining them. Images, such as "concrete

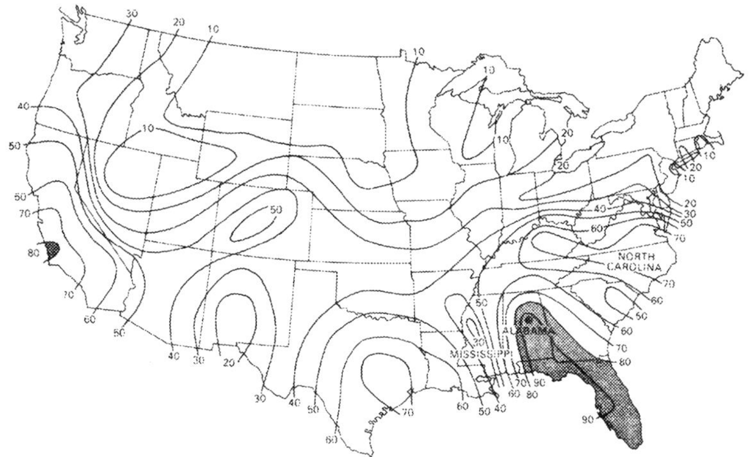

Figure 2. The Mental Map of the US from Alabama (Gould and White 60)

jungle," provoke stories, such as the ones shown on *New York Law*, which then recondense into images and maps ("Don't go there!"), and so on, in a cyclical process.

Although the maps shown here seem to depict the nation-state of the United States, that may not be so. They may in fact depict only regions or places; they say little about the United States as a whole, other than that identification with its various states and regions varies with location. The isobars also reveal that mental maps create boundaries and borders. These may be physical, ethical, behavioral, or linguistic, according to the kinds of stories that go into their making, as explained above. In literary texts, these boundaries become the distinction between Self and Other, who also become indispensable to each other in order to keep the boundary intact. Common placeholders for this distinction in the modern period are: the national versus the foreign; and (as in these examples) the national versus the regional.

Diataxis

Literary texts as mental maps are distinguished by an intense foregrounding of what Hayden White calls their "diataxis," which is a relational term for the process of transforming description into plot, and vice versa:

> discourse must be analyzed on three levels: that of the description (mimesis) of the "data" found in the field of inquiry being investigated or marked out for analysis; that of the argument or narrative (diegesis) running alongside of or interspersed with the descriptive materials; and that on which the combination of these previous two levels is effected (diataxis). The rules which crystallize on this last, or diatactical, level of discourse determine possible objects of discourse, the ways in which description and argument are to be combined, the phases through which the discourse must pass in the process of earning its right of closure, and the modality of the metalogic used to link up the conclusion of the discourse with its inaugurating gestures. (4-5)

We see here the familiar triadic relationship between terms, so that the meaning of each depends upon its relationship with the others. White's language implies that the movement is always in one direction: diataxis—which means literally "movement through"—simply helps us move from mimesis to diegesis. In fact, however, to repeat a point made earlier in explaining Gould and White's maps, the movement between the three levels is continual and cyclical. As the term "mimesis" implies, all three levels of discourse are (in my case) linguistic constructions. This is not a positivist or Darwinian theory of the shaping of narrative forms by the raw data of lived environments, although we will encounter such views in some of the authors under examination. The mimetic appropriation of landscape is constructed, and imbued with intentionality and ideology, to the same extent that narrative is. The difference is between synchronic and diachronic, spatial and temporal discursive modalities.

Simon Schama's book, *Landscape and Memory*, addresses exactly White's dialectic, between the use of myth to define landscape and the ability of landscape to provide continuity to cultural myths and traditions. He writes, "our entire landscape tradition is the product of shared culture . . . a tradition built from a rich deposit of myths, memories, and obsessions" (14). Schama's approach, derived from the phenomenology of Gaston Bachelard, relates specific cultural images of woods to the single image of the tree of life, and cultural images of bodies of water to the *acqua vita*. His approach to cultural memory is that of landscape archetypes as schemata into which one may place various and conflicting histories. In some sense, then, the title of his book refers to landscape as an artistic genre rather than as an environment. Schama's landscapes are broadly conceived: forests; rivers; and mountains. To give one example, he takes as the subject of the first part of his book several European forests: Bialowieza in Poland; the Teutoburger in Germany; and Robin Hood's "Sherwood" forest in England. Schama notes in each case the intersecting histories of the forest as physical habitat and as repository of cultural memory. These two histories never overlap completely, and at times they directly contradict each other, as in the Tudor period when "the Crown presented itself as the custodian of the old, free, greenwood [while] it was busy realizing its economic assets" (153). Schama is the Northrop Frye of landscape studies; his book amounts to an *Anatomy of Landscape*. As with Frye, the trees are obscured somewhat by the forest of Jungian archetypes. Rather than subsuming the variety of landscapes encountered in this text under a limited category of labels, my focus is on the shaping of particular diataxes through processes of cultural production.

A small example

Only a small portion of "mapping" activity occurs as formal, graphic representation. Much of it occurs in the stories people tell each other, in the artworks and crafts produced in particular areas, and, as in this book, in the literature that transforms regional and national mimesis into diegesis. Let us examine the mimesis-diataxis-diegesis triad with a literary example that takes on cartographic functions. Take, for an example, the diataxis of Robert Johnson's much imitated blues classic, "I Believe I'll Dust My Broom," which he recorded in 1936. That Johnson sings a geography of the Mississippi Delta can be seen from the appearance of photo essays in journals such as *Natural History*, which describe his music in relation to the region. This particular song moves from more realistic to ever more fantastic heterotopias. Things begin mundanely enough: "I'm gon' get up in the mornin' / I believe I'll dust my broom. / Girlfriend, the black man you've been lovin' / Girlfriend, can get my room" ("Dust My Broom"). So far, this is a simple leave-taking song, prevalent in the blues tradition; its diataxis follows the familiar pattern of theme and improvisation. The singer's lament begins modestly, with a realistic invocation of the geography of the delta, and of the life of a traveling musician: "Girlfriend, the black man you've been lovin' . . . can get my room." While the context of the whole song, as well as of the blues tradition, makes it clear that the end of a relationship is causing this move, the line could also be read as a straightforward statement of the life of an itinerant musician, of a "steady rolling man," as another Johnson song puts it, but also of the general movement of African Americans driven by economic and political necessity. The latter went from South to North, a pattern Johnson was undoubtedly familiar with, and that the Blues itself would follow to Chicago, electrifying itself in the process, as in Elmore James's famous rendition of this song. (Johnson's own best-selling recording was "Terraplane Blues.") It's here today and gone tomorrow, from East Monroe to West Helena—two Arkansas towns located a few miles apart from each other: "I'm gon' write a letter / telephone every town I know / If I can't find her in West Helena / she must be in East Monroe, I know." In my field research in Mississippi (see "From Oxford to Jefferson" below), I drove to West Helena, which celebrates itself as a Blues center. I saw the sign for Monroe, but there is in fact no East Monroe, Johnson made it up to keep his map symmetrical. Johnson implies a realistic social architecture: the itinerant musician always gives up a room, not a house. If the departing ex-lover invokes the itinerant musician, the itinerant musician invokes, in his turn, an entire society of transients, the landless sharecropper class—in 1920 blacks made up three-quarters of the Delta population but owned less than three percent of the land.

The last stanza of the song raises the stakes, exploding into a fantastic geography—the singer's voice trails a bit behind the guitar line here, as if burdened by the imaginative leap involved: "I'm 'on' call up Chiney / see is my good girl over there / 'f I can't find her on Philippine's Island / she must be in Ethiopia somewhere." The stanza both refers to the real world, and also "signifies" upon the first and second stanzas. The whole song, like the genre of blues in general, is an example of signifying in which the complaints addressed to a specific woman are really addressed to society in general, fulfilling signifyin(g)'s definition of "arriving at direction through indirection" (Gates 44-88). Johnson parodies his own earlier hopes of finding his girlfriend in the impossible, fantastic geography of telephoning China and going to the Philippines. The mention of Ethiopia may be a bitter reference to the condescending term "Ethiopian" used as a name for African Americans. James Kimble

Vardaman, a populist Mississippi politician, used the term in his racist speeches, saying, for example, that "We would be justified in slaughtering every Ethiop on the earth to preserve unsullied the honor of one Caucasian home" (Kirwan 146-47). Johnson tropes the white usage of the term by "laying bare" its solecism, exposing it by inventing his own for the Philippines, which appears in his usage to belong to a female Philip. He thus signifies on an imperialism which, although it operates in far-off lands, is an arm of the same system that keeps Johnson "in his place," bouncing around the Delta. Johnson's geographical references are culled from news reports of the Japanese in Manchuria, of American troops in the Philippines, and from the Italian invasion of Ethiopia. His references aggressively turn the restricted geography of black migration inside out, as it were, where it becomes identical with the unlimited travel and surveillance of European exploration and possession. In the midst of the other geographical impossibilities—cultural, economic, and technological impossibilities—Ethiopia becomes a bitter possibility. The final stanza rejects the previous one, substituting its belief and hope with despair, or alternatively, rejecting its realism for a certain idealism. Mixed with the bitterness, after all, is a geographic expansiveness that suddenly stretches the thirty miles of Arkansas backroads into a trip around the world.

Johnson's lyric is a rhetorical or imaginative exercise, but surely not a map? More than one geographer urges us to accept the position that "all maps are rhetorical texts," and to dismantle "the arbitrary dualism between . . . modes of 'artistic' and 'scientific' representation as they are found in maps" (Harley 242). Ann Buttimer takes up a similar stance: "When a geographer wishes to make sense of that vast panorama of fact and fiction, pattern and event, on the surface of the earth, what goes on in his consciousness? How does a geographic sense of reality emerge, and how does it distinguish itself from that of the geologist, poet, painter, or historian?" This geographer answers her own question with the phrase "chorographic awareness," which can "best be understood as a poetic rather than a calculative exercise" (12). I will not be exploring the degree to which geography might be deemed a poetic task; rather, I examine situations in which poetry ends up performing geography.

The cultural production of mental maps

From time immemorial, as Kevin Lynch tells us, and as we may conclude from some of the preceding argument, place and landscape have performed a social function: "The named environment, familiar to all, furnishes material for common memories and symbols which bind the group together and allow them to communicate with one another. The landscape serves as a vast mnemonic system for the retention of group history and ideals" (*Image* 126). In this sense it is true or, to speak in the manner of Huck Finn, it at least ain't no lie that Indiana is pink, because that is the way Indiana's reality has been constructed on the map Huck saw.

In comparison with Johnson's song, certain maps seem to us rhetorically neutral because they are produced "scientifically." That is, the mimesis appears to control or suppress the diegesis—as opposed to the Bostonian's map of the United States, where the diegesis distorts the mimesis. We may distinguish such maps ("maps" in our daily usage of the word) from Johnson's talking map, in which the mapping function is not separated from the actual living processes of human beings in their environments. The former method corresponds with Martin Heidegger's analysis of science as a particular *mathesis* that subjects natural phenomena to a schema ("Entwurf") in which things lose their individual qualities in the face of the natural laws governing them: "The mathematical is, as *mente concipere*, a schema of things which seems to move beyond ["hinwegspringen über"] their thingness. The schema itself opens a field in which the things, i.e. the facts, reveal themselves. . . . The kind of questioning and cognitive definition of nature is no longer regulated by means of traditional opinions and concepts. Bodies have no hidden qualities, powers, or capabilities. Natural bodies are merely what they show themselves to be in the domain of the schema" (92-93). Thus, the Greek model of the universe as a series of concentric shells holding up the astral bodies falls outside the world picture of modern science not because their model was intuitional rather than a hypothesis constructed from observational data, but rather because within it each body, like a work of art, had its own unique place and role to play, its own *aretè*, rather than fulfilling a single law of nature that treats all bodies alike. Edward Casey has written a detailed philosophical history of the transition from place to space that confirms and fills in the details of Heidegger's sketch. In philosophy, the ascendancy of space over place occurred during the early modern period in the writing of "Descartes and Gassendi and Newton, Locke and Leibniz and Kant himself in his early years. All presume and promote the supremacy of space; none hesitates to submerge places (properly plural) into space (only singular)" (*The Fate of Place* 193). This principle of an all-encompassing universality and neutrality inhabits neither Johnson's song, nor any of the true imaginary places we will be visiting in this book. Many of the texts considered, however, themselves fight out the battle between abstract, spatialized conceptions of the world, and a sense of the uniqueness (which means, in this context, the unmappability) of the true imaginary places they create.

Places engender stories that explain these places and make us more aware of them, and so on. As Keith Basso puts it, as a "variety of experience, sense of place also represents a culling of experience. It is what has accrued—and never stops accruing—from lives spent sensing places" (83). Basso gives an example from the Arizona landscape. When a young Apache cowboy recovering from a bout of wildness asks to be taken back on by the elder cowboys, they respond with: "So! You've returned from Trail Goes Down Between Two Hills." This seeming reference to a true place, whose topography resembles that of a "notch," is really an allusion to a group of stories associated with that place, concerning Old Man Owl and two beautiful women who tormented him. Each story uses the features of the landscape; for example, the women had set themselves up on the two hills and taken turns tempting Old Man Owl. He went back and forth between them until he was exhausted. The elders situate their younger colleague morally and ethically by situating him in place, and they do the latter by "mapping the invisible landscape" of telluric folklore, to use the evocative title of Kent C. Ryden's study. In this anecdote, story helps define place, but place itself assumes a didactic function by telling a story. Mary Layoun has done similar collecting for Cypriot, Greek, and Palestinian identities in *Wedded to the Land?* Not coincidentally, the situations Layoun studies involve mass displacements and governments-in-exile, which perhaps shifts the cycle in favor of narrative as a means of constructing nationhood, since the territorial aspect becomes contested and fragmented and is less portable.

To clarify further the process by which information is transformed from perception of place into narrative, from uniquely private into communally public, I reproduce below Richard Johnson's graphic of the cycle that outlines the pathway by which individual, private, lived experience (at the bottom) undergoes a production process and becomes public. In this case, simply replace "texts" in box 2 by "mental maps":

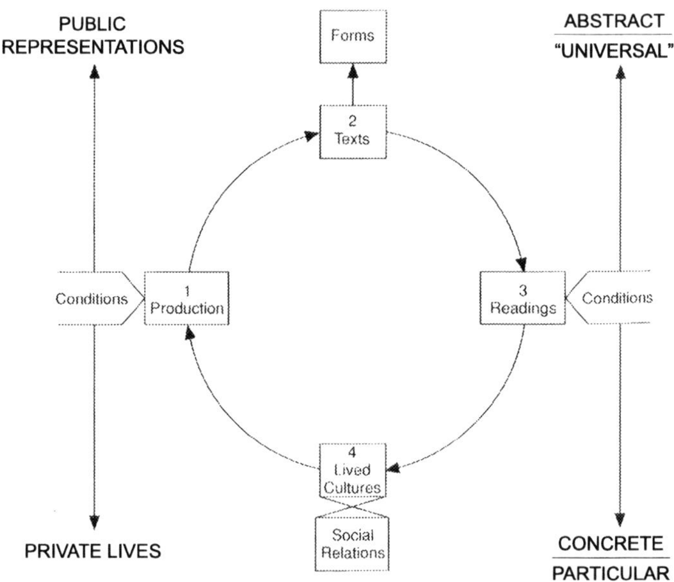

Figure 3. Diagram from Richard Johnson, "What is Cultural Studies Anyway" (47)

On the left-hand side of the circle is "production," that is, the conversion of private perceptions and personal stories into public forms. This conversion can be considered an aspect of diataxis. It so happens that this graphic was designed by Richard Johnson as an explanation of the object of cultural studies, which is to study the "processes . . . intrinsic to cultural circuits under modern social conditions, and that . . . are produced by, and are productive of, relations of power" (49). The texts I investigate are mental maps (or rather, I investigate them as mental maps) and the form of mental map I investigate is the literary text, broadly defined—as opposed to Basso, who uses other forms of cultural memory. The cultural studies approach involves an attempt to grasp and understand as many points of the circuit as possible (as opposed, for example, to a purely formal analysis that would intersect with the top of the circle, to a psychoanalytic approach that would bring to light culture as lived, private lives, the particular, and other elements from the bottom, and so forth).

The diagram implies the important role that social relations and "conditions" have on the production of maps of any kind. In his study of the political function of mapping, Mark Monmonier notes that "folks who are routinely suspicious of written text equate maps with fact and fail to realize that no map is capable of including all information or telling all possible stories. In fact, the process of mapping requires cartographers to limit content in order to create a readable map and so allows them to manipulate their audience with the information they choose to include. This combination of power and subjectivity has repeatedly put maps at the center of controversy" (1). As other geographers note, "the detail and complexity of reality is selected, simplified, and then emphasized, so as to portray only what the map maker believes to be the essence of the map referent space, as defined by the purpose of the map" (Muehrcke 333). In the Middle Ages, English cartographers felt justified in drawing their *mappae mundi* from what looks like a polar perspective, with Jerusalem in the center, and the land masses of Europe, Asia, and Africa arranged more or less equally around it. While Columbus insisted that he had not discovered a new continent, but simply landed at the eastern shores of Asia, mapmakers insisted on putting America to the left-hand side of the map, and Asia to the right-hand side. Their activity did not reproduce existent data concerning the independence of the American continents, but played a role in shaping the mental image of America as a landmass in and for itself. Eviatar Zerubavel notes that this placement was the equivalent of the rhetorical arrangement of narration. In each case, boundaries must be set in order to determine beginnings and endings, silently eliminating what doesn't "count": "As evident, for example, from the way groups manipulate the temporal boundaries of historical narratives, the act of drawing a boundary (such as deciding what will constitute the beginning or the end of a narrative) often obfuscates the very existence of whatever lies beyond it. By the same token, by placing America on the far left side of their world maps (thereby establishing a lasting cartographic convention that has in fact prevailed to this day), early-sixteenth-century mapmakers clearly helped promote in Europe's mind the absolute separateness of the New World from the Orient" (Zerubavel 74). From this point on, European maps of the world tended to place Europe in the center, with the Americas on the left and Asia on the right. Had nations been mapped before the idea of the nation-state became prominent, a world political map would have resembled a painting by expressionist Oskar Kokoschka, a "riot of diverse points of color" (Gellner 139). Political maps generally are divided into nation-states, and resemble paintings by Amedeo Modigliani or Piet Mondrian, in which "neat flat surfaces are clearly separated from each other" (Gellner 139-

40). Current chorographic techniques are not geared toward the representation of situations of sub-state nationalism, borderlands, or diaspora that compete with more straightforward ideas of nation-state in today's world. (It is interesting to speculate on the forms of chorography that might emerge in the globalized future, with work being done by Mark Nunes, Laurie Piper, and others on this topic that intersects with information technology and cybernetics and is beyond the scope of this book.) Among the many types of maps, mental and otherwise, literary ones distinguish themselves through the rich detail they provide on private perception, the fluidity of movement in them from mimesis to diegesis and vice versa, and their foregrounding of diataxis. The "protagonists in motion" of these literary works, through their traversal of space, topography, and social networks, confront and (re)create their mental maps of nation and region.

Literature as mental map

This book focuses on literary texts as a valuable source of "discursive territoriality" (Häkli) contributing to the construction or deconstruction of national, local, or regional identity. Literary texts form an exception to the rationalizing processes of modern cartography. Yet, in his thorough sifting of geographers' approaches to the novel, Marc Brosseau notes the lack of methods appropriate to the literary text. By default, there is a wish "to ask of fiction that it become an instrument for [geographers'] basic data collection" (31). Brosseau attempts to construct a dialogue between geographic and novelistic discourse, but stops short of allowing literature a role in the unconcealment of the hyphen in "geo-graphy," that is, the interrogation of what might constitute the boundaries of the writing of the earth. As Nigel Thrift indicates, along with the visual arts, literature (in the broad sense) has been the most complex and seductive form of geographical simulation throughout Western history: "places have meanings and meanings are always produced, never simply expressed, as part of a wider process of cultural creation. Literature is one way in which such meanings are produced within a culture and ascribed to place, just as place is often appropriated to produce meanings in literature" (21). Anderson, for example, argues that the novel, which he reduces to the picaresque genre—"the movement of a solitary [picaresque] hero through a sociological landscape"—is powered by this "national imagination," but clearly this is a dialectical relationship (30). Perhaps the most famous example of a wandering, solitary hero is Don Quixote, who has since become an abiding symbol of Spanish cultural identity. The combination of nation, chorography, and mental maps under the rubric of heterotopia furnishes a critical template for comparing writing as disparate as Johann Wolfgang von Goethe's presentation of Italy to the Germans, Nikolai Gogol and Ivan Turgenev's competing Russian geographies, Euclides da Cunha's ambivalent but impassioned nomination of the *sertanejo* to the position of "bedrock of the Brazilian race," William Faulkner's and José Lins do Rego's cultural memories of the Northern and Eastern Caribbean, respectively, José Maria Arguedas's novelistic ethnogeographies of Andean culture, Juan Benet's

"Région" as both metaphor and metonym for Francoist Spain, and the "u-topian" and deconstructive desert landscapes of Mary Austin, Nicole Brossard, and Joy Harjo. These texts have been chosen so as to provide a chronological and ideological range within the modern period, beginning with the deliberate prenationalism of Goethe, and ending with the pronounced postnationalism of Nicole Brossard and Rolf Dieter Brinkmann. In between, we see a variety of conditions and attitudes, from Gogol and Turgenev's attempts to find their way in the uncertainties of the Russian Empire, to Euclides da Cunha's attempts to fence off the Brazilian Republic from its backlands, to Arguedas's attempt to narratively transculturate the "Piruvian" nation, to the open embracing of regionalist exceptionalism by William Faulkner, José Lins do Rego, and Mary Austin.

Like Otmar Ette's study, *Literature in Motion*, this book also "interrogates literature regarding the evolution of spatial concepts, which . . . were meaningful for the last quarter of the second millennium and that emerge above an aesthetic and 'spatiality' of the modern" (9). Ette's is the first full-length comparison of the effects of motion and space—as interrelated rather than independent phenomena, and as geopolitical rather than physiological processes—on the development of modern literature. One of the most important historical formations deposited by these processes, as Ette points out, is the two-edged development of "American thought": thought in the Americas, as a hybridization of European intellectual traditions with telluric concerns; and thought of the Americas by Europeans whose world view changed drastically after 1492. My study for the most part ignores the important aspect of return from America into Europe that Ette traces so well (an exception will be my discussion of the influence of Euclides da Cunha on Juan Benet); it resembles instead a colonist, beginning with European authors and geographies and embarking at some point for the Americas, never to return, swallowed by the immensity and splendor of the earth.

As its title indicates, Ette's book focuses on travel literature, and especially on travel beyond national boundaries. Although movement is a necessary component of "inspecting the boundaries" of one's mental map of local, national, and regional spaces, transgression of those boundaries is not. Of the works I examine, only Goethe's would be considered as belonging to the category of "travel literature." On the contrary, many of the works—most notably, Gogol's *Dead Souls*, Turgenev's *Notes of a Hunter*, Austin's *Land of Little Rain*, and Brossard's *Mauve Desert*—are characterized by an intense and often repetitive movement of the protagonists that does not amount to travel. The title of Benet's *Return to Region* sums up the repetition compulsion and eternal recurrence that characterize these texts. Elsewhere, stasis seems more appropriate, as in the illustrated prose poems of Harjo and Stephen Strom's *Secrets from the Center of the Earth*, where the stillness of the photographic image seems necessary for the drawing of diegesis out of mimesis. Readers familiar with these works may have noted my solecism in claiming the status of fiction for works of autobiography (Goethe), history (Da Cunha), and essay (Austin). My argument is admittedly circular: my hypothesis that these authors are constructing heterotopias causes me to call their work fictional. Secondly, however, in none of the cases mentioned am I the first to claim fictional status for the work in question. Thirdly, the rubric of fiction is meant to emphasize the shared foregrounding of diataxis in these works, rather than to dispute that this or that reported event "really happened." Other readers undoubtedly will be supplying their own favorite examples that I have missed; I have selected texts that have a particular resonance within their national cultures, assembled a range of them so as to give both historical (early modern to contemporary) dimension and geographic range (Europe to the Americas) to this study, and compared them with one another wherever possible. By interpreting and comparing these fictions, I engage in what Foucault calls heterotopology, a "simultaneously mythic and real contestation of the space in which we live" (24).

Raymond Williams's idea of the "knowable community," presented in his *The Country and the City* provides an example from literary criticism of the difference between chorographic and mental mapping. Edward Saïd, in acknowledging his debt to Williams, summarizes the latter's contribution as "demonstrat[ing] that literary and cultural forms . . . derive some of their aesthetic rationale from changes taking place in the geography or landscape as the result of a social contest" ("Invention" 248). The English landscape as represented in literature is not physical but imaginary, a series of mental maps, in which the contour lines chart the degree of representability of "real" landscape, architecture, and classes of people. Williams's brief description of Jane Austen's mental map of semirural England is particularly apt: "Neighbours in Jane Austen are not the people actually living nearby; they are the people living a little less nearby who, in social recognition, can be visited. What she sees across the land is a network of propertied houses and families, and through the holes of this tightly drawn mesh most actual people are simply not seen" (166). Williams terms this a "knowable community." We know what Austen "saw" in the landscape through what she chose to describe and to narrate. Again, there is nothing unusual in Austen's exclusions; it is what mapping is all about. We use highway maps to locate highways, hydrographic maps to locate rivers and streams, and political maps to locate countries. The surprise is to discover this process at work in the heart of a novel, and beyond that, to suggest that such mapping constitutes one of the functions of modern fiction.

In its small-scale approaches, as Yi-fu Tuan points out, contemporary geography approaches the feeling of place that literature has always provided: "Modern scientific environmentalism is modest in its claims. No longer does it pretend to explain national traits, whole cultures and civilizations. Its contributions now lie in the attempt to relate physical characteristics of the environment to human behavior within well-defined contexts. At one end of the scale the architectural psychologist tries to show how the slope of the ceiling in a room influences the friendliness of its occupants; at the other the urban sociologist seeks to demonstrate how the spatial and even social behavior of the urban populace is influenced by the physical structuring of the neighborhood (*Man and Nature* 28). The conveyance of this interiority has traditionally been the job of literature. As Ronald Bordessa has suggested, a "true synthesis of geography and literature would produce an entirely new approach to meaning" (273). Readers will find several attempts at such synthesis in the chapters of this book; may they also find new meanings for some familiar and valued literary artifacts. Among geographers, however, only Brosseau has ventured fully into the realm of literature by daring to coin the term "novel-geographers" ("romans-géographes"). Note the double noun: these are not geographical novels, he claims, but novel-geographers that truly investigate people's experience of space. Brosseau's comparative analysis of works by John Dos Passos, Julien Gracq, Claude Simon, and Patrick Süskind can be said to break the path for my own investigation, which, however, adds a more stringent comparative dimension to the deliberate march through history and across the Atlantic.

Exemplarity and comparison

Some of the authors treated in this study could be usefully related to Saïd's concept of an imperialist geography: Goethe's travel to Italy follows the path of centuries of German *Kaisers*; Gogol mythologizes a Russian empire unified in its one language even as he shows the lostness and disorientation of its subjects; and da Cunha primitivizes the Brazilian backlanders in full realization that to do so is to recolonize Brazil along European models of behavior and thought. In general, Saïd's attention to images of the Other as essential for European self-understanding, and his persistent questioning of "what is at stake?" in acts of representation are useful analytical tools even in the context of the "domestic geographies" I have chosen for this study. However, neither landscapes nor their fictional embodiments can be reduced to the intentions—be those imperialist, nationalist, or deconstructionist—of their "creators," as Jouni Häkli cautions: "the concept of discursive landscape does not denote a closed, clearly defined and instrumentally applicable device of social control. The specific knowledges, images, and symbols giving shape to national identity cannot be fully reduced to the direct motivations and aspirations of the elites who produced them. Rather, it should be noted that all collective identities are subject to historical transformation, and because they arise from different social bases, they consist of various contestable and contradictory elements" (124-26).

In landscape as in literature, the roles of reader and interpreter—on the right side of Johnson's circle—can easily frustrate any "authorial" intentions: "The social-life-as-text metaphor is easily applicable to landscape because it too is a social and cultural production. Thus a landscape ... becomes detached from the intentions of its original authors, and in terms of social and psychological impact and material consequences the various readings of landscapes matter more than any authorial intentions. In addition, the landscape has an importance beyond the initial situation for which it was constructed, addressing a potentially wide range of readers" (Barnes and Duncan 6). In fact, however—with notable exceptions—the texts examined in the following chapters were written by authors fully aware of the implications of national or regional identification. Their texts reveal the "contestable and contradictory elements" at the basis of all heterotopias. The authorial strategies and perspectives comparatively explored in this book point to alternative notions of mapping as intentional, divisive, and mysterious.

Rather than examine chorographic representations that divide the world into neat categories and objects of empire, then, this book concentrates on a process of fractalization, in which the facticity of landscape and topography reticulate into infinite pockets of uniqueness that "are not down on any map." The key to this complexity lies in the self-reflective focus on diataxis in literary texts. This approach corresponds to the refinement Homi Bhabha has contributed to Anderson's idea of "imagined community." As Bhabha has pointed out, the location of nationality within discursive practice means that such imagined communities are, like language itself, inherently unstable and heterogeneous. Anderson "did not quite take into account the complexity of the functioning of the sign in its incessant and always-renewed structuring of meaning" (Chanady x). Bhabha's work deconstructs the assumed bird's-eye perspective implied in Anderson's term, as when the former writes that "to encounter the nation as it is written displays a temporality of culture and social consciousness more in tune with the partial, overdetermined process by which textual meaning is produced through the articulation of difference in language; more in keeping with the problem of closure which plays enigmatically in the discourse of the sign" (Bhabha 2). Take, for example, the concept of a common territory as one of the key elements of nationhood. Positivist approaches would emphasize the actual processes of mapping, defining, negotiating, maintaining, and defending national territories. Anderson, in contrast, would emphasize the embodiment of territory in national consciousness.

Just as some studies have followed Saïd's lead in discussing the role of cartography in domination and possession, others have addressed the contradictions and complexities involved in the mapping of heterotopias. In the conclusion to his comparison of cartographic themes in contemporary Canadian and Australian fiction, Graham Huggan attributes a political project to such fictions: a "connection can be made between the symbolic destruction and/or reconfiguration of the national map and the ongoing project of cultural re-evaluation. In the former 'settler colonies' of Australia and Canada, creative writers and critics have participated in a project of cultural decolonization characterized by the gradual displacement/replacement of a 'cartography of exile,' based on the self-privileging value systems of European colonialism, by a 'cartography of difference,' constructed on principles of cultural diversity that are more appropriate to the heterogeneous nature of post-colonial societies/cultures" (147-48). Moving through several centuries of time and a number of languages and cultures, this comparative study adds complexity to Huggans's binary opposition—which is essentially one between politically motivated cartography and subversive literary mental maps. Goethe's Italian journey maps nearly everything except political boundaries. The fictions of Gogol and Turgenev map Russia as a discontinuous, fractalized political nonunit. As Huggan points out, the regionalist approach that shapes the narratives of da Cunha, Lins do Rego, Faulkner, and Benet presents significant problems for boundary making. Finally, the fictions of Austin, Joy Harjo, and Brossard challenge nationalism self-consciously with the "cultural decolonization" of desert regions.

It will be noted that the heterotopias treated in this book are mostly nonurban, although by the terms of comparative methodology I should also be treating localities side-by-side with larger spaces. Some investigators do not feel that descriptions of urban locales attempt the identification of reader with larger political units that one finds in the texts under consideration here. According to Burton Pike, the themes of urban landscapes are variations on "paved solitude" (xii). Margaret Atwood has addressed the difference between urban and rural heterotopias in relation to Canadian identity and to the production of Canadian literature. In designing a series of lectures on the latter topic, published under the title, *Strange Things*, Atwood felt compelled to make a choice between addressing the urban experience of the majority of the Canadian population, and the "Malevolent North" (analogous to Gould and White's "Southern Trough") that figures so prominently in its literature. Her choice of the latter as source of more evocative mental maps of Canada than Toronto or Ottawa provides an excellent example of heterotopia—a familiar yet distorted image of where Canadians are "from." Atwood was challenged by a compatriot, in flesh and blood, to defend her choice, and came up with a pragmatic answer:

> I was interviewed by a young man from Canada. . . . He told me that he had a friend—also Canadian—who was concerned about the subject matter I was discussing. This friend felt that I should not be talking about the North, or the wilderness, or snow, or bears, or cannibalism, or any of that. He felt that these were things of the past, and that I would give the English a wrong idea about how most Canadians were spending their time these days. What then—I asked—did this young man think I should be discussing? "The literature of urban life," was the reply . . . I failed to say that the right idea could often be right from a sociological point of view, but was not necessarily right from a literary one. Given a choice between a morning spent in the doughnut shop and a little cannibalism, which would you take—to read about, that is? (Atwood 5)

The zeugma of Atwood's last sentence, linking place ("the doughnut shop") with event ("cannibalism") provides yet another example of the open boundary between description and narration. Reality and ideology, Atwood reminds us, stand in a dialectical relationship. The relative ability of a cityscape or landscape to become a heterotopia does not depend upon a plebiscite. Nor, on the other hand, does it depend upon the genius of a single, towering author—although it might look that way in the cases of the authors considered in this study.

Chapter Two

Landscapes of Nation in Goethe's *Italienische Reise* and its Counter-Narratives

Since the publication of Goethe's *Italian Journey* (*Italienische Reise*; *IR*) in 1816/17, the Italian landscape has been extraordinarily important—more important, perhaps, than the German landscape—for German intellectuals' understanding of themselves, of history, and of culture. The heterogeneity of the text that came to be known as Goethe's *Italian Journey* derives in part from a complexity of its author's position vis-à-vis the two cultures in question—the Italian one which is its subject, and the German one which is its object—a complexity that transcends the aphoristic summations of his twentieth-century avatars. Goethe's cultural and political conservatism are well known; this chapter examines the particular use he makes of landscape in the *IR* as a symptom of his prenational political perspectives. Given that both Germany and Italy were "nations without statehood" in the late eighteenth and early nineteenth centuries (i.e., were united linguistically and culturally, but not politically), these linkages are implicit, and the bringing of them to light requires ancillary texts such as the Italian *Tagebuch* and some reception history of the *IR*, as well as counter-narratives by Alessandro Bárbero and Rolf Dieter Brinkmann, who self-consciously echo themes of the *Italienische Reise* in the context not of full-blown nationalism, but of globalization and postnationalist perspectives. This examination of a complex of interwoven texts will exemplify the cultural cycle that produces particular landscapes out of the dialectic between private experience and public discourse.

The complicated nexus of concerns in *IR* is reflected in its history of composition and publication. While in Italy, Goethe wrote letters home continually and also kept a notebook and various other annotations. Upon his return to Weimar, he published a few scattered writings, such as the description of Roman carnival, in various journals. He eventually put together the bulk of the *IR* from these heterogeneous writings. Melitta Gerhard has pointed out the systematic nature of the changes Goethe wrought in this material. Goethe systematically eliminated the dialogic aspects of the letters, for example, and turned their singular addressees into an indefinite, northern "Leserkreis" ("readership") for his dispatches from Italy, with

Figure 4. Italy resized according to the number of pages Goethe published on each region

the result that "everything non-Italian is now drawn into the projection of the Italian atmosphere" (40). Corresponding to Goethe's symbolic appropriation of geodesic vocabulary into the cultural realm with which he is concerned is his transformation of the material of his own writings for the preparation of his narrative. The effect is to make the traveler reporting his Italian experience into a fictional character equally different from the thirty-seven-year-old who made the trip and the sixty-seven-year-old who published the text. There are at least two names for this process: Lilian Furst has called it the "intercalation of the writing self" in the *IR* (129), while Jean Lacoste considers it a "participative sociology" (58). Logically, there are two methods for intercalation: spatial and temporal. That is, Goethe can (verbally) paint a landscape or other scene in which he himself appears; or, he can tell a story in which he himself enacts a role. So-called travel narratives nearly always present a composite of narration and description, mixed in some proportion. Like modulation between keys in common-practice music, the rate and method of change from one mode to the other becomes part of the work's significance. Thus, we have four possible dispositions determined by the two binary choices: narrate versus describe; and take on the role of an engaged versus a distanced narrator or descriptor. These choices are constantly changing and determine much of the texture of *IR* in ways to be analyzed here.

Goethean diataxis

Goethe uses the mimesis/diegesis distinction to define the parameters of imaginative geographical inquiry in his book: "we are driven back and forth between nature and cultural events" (228) ("Und so wird man zwischen Natur und Völkerereignissen hin und wieder getrieben"; 187). Nature is mimesis, to be described, while cultural events are to be narrated and belong to diegesis. Goethean diataxis, then, is being "driven back and forth" between the two as their interpretant.

As a travel narrative, Goethe's *IR* has a variety of mimeses and diegeses. Landscape must tell the story of Italy, of Germany, but also of Goethe himself. As Werner Gephart points out, the journeying self is constituted out of the conjunction of a variety of social domains, that of departure, that of transit, and that of entry (107). What is considered one's own is compared with the "foreign." In Goethe's case, the landscape of departure is returned to in accordance with Hans-Georg Gadamer's idea that "the essence of *Bildung* is not alienation as such, but rather the return home to oneself" ("nicht die Entfremdung als solche, sondern die Heimkehr zu sich, die freilich Entfremdung voraussetzt, macht das Wesen der Bildung aus"; 11-12). *Bildung* can be defined as the incorporation of the foreign into oneself, which, to use Susan Bernstein's formulation, "involve[s] a moment of appropriation of sensuous material into a meaningful structure, or one of the incarnation of meaning in a sensuous body" (1016).

Bildung will also become Goethe's codeword for the political and social renewal of Germany upon his return from the south. Rejecting the radical solutions of the French Revolution, Goethe stressed that Germans would only become capable of forming a republic once they had become educated. In 1795 he outlined a dialectical view of the relationship between national authors and the nations they represent: a national author is someone who actively participates in the ability of a nation to conceive itself and its goals. At the same time, a national author must find his materials in the thoughts and actions of his compatriots: "If the writer finds in the history of his nation great events which together with their consequences form a harmonious and significant whole; if his countrymen exhibit nobility in their attitudes, depth in their feelings, and strength and consistence in their actions . . . if his native country has attained a high level of culture, thus facilitating his own educational process" ("Response to a Literary Rabble-Rouser" 190; "Wenn er in der Geschichte seiner Nation große Begebenheiten und ihre Folgen in einer glücklichen und bedeutenden Einheit vorfindet; wenn er in den Gesinnungen seiner Landsleute Größe, in ihren Empfindugen Tiefe und in ihren Handlungen Stärke und Konsequenz nicht vermißt . . . wenn er seine Nation auf einem hohen Grade der Kultur findet, so daß ihm seine eigene Bildung leicht wird"; "Literarischer Sanskulottismus" 240). This list of prerequisites amounts to Goethe's ultimatum to Germany, and his consequent resignation from any attempt at becoming Germany's national author, since in Goethe's view, "Germans were unwilling to perfect themselves; they lacked *Bildung* and it was not apparent that they were about to acquire it. . . . Goethe clearly generalized and emphasized that the Germans were unable 'sich zu bilden,' men and women, and that this lack of 'Bildung' was one of their major national faults" (Koepke 92). Hence the turning away from politics and the lack of pro-German nationalistic statements in Goethe's post-Italian works and letters, as noted by Wolfgang Rothe (216-36). This unformedness of the Germans contrasted with the Italians, who regardless of social class had been shaped by—indeed, who lived their lives surrounded by—their classical heritage, at least in Goethe's way of seeing as transmitted in *IR*. Near the end of his life, Goethe ascribed the dialectic of *Bildung* to the Napoleonic wars and named its result as *Weltliteratur*, thus projecting his own biographical experience onto the larger dimension of cultural and historical developments: "For some time there has been talk of world literature, and properly so: for it is indeed evident that all the nations, thrown together at random by terrible wars . . . could not help realizing that they had been subject to foreign influences, had absorbed them and occasionally become aware of intellectual needs previously unknown" ("World Literature" 228; unless indicated otherwise, all translations are mine; "Es ist schon einige Zeit von einer allgemeinen Weltliteratur die Rede, und zwar nicht mit Unrecht: denn die sämtlichen Nationen, in den fürchterlichen Kriegen durcheinander geschüttelt, sodann wieder auf sich selbst einzeln zurückgeführt, hatten zu bemerken, daß sie manches Fremde gewahr worden, in sich aufgenommen, bisher unbekannte geistige Bedürfnisse hie und da empfunden"; "Einleitung zu Th. Carlyle" 364). Perhaps because it was so distasteful to him, Goethe does not mention here the other side of the dialectic, the rise of German nationalism as a reaction to Napoleonic imperialism. Goethe's admission here of the rearrangement of Europe through the Napoleonic wars contrasts strangely with the narrative strategy of *IR*. Goethe doubled the *Verfremdungseffekt* of the *IR* by publishing it at such a distance—across the turn of a century and a revolution—from when the travel took place. Goethe had seen his journey to Italy as a necessary beginning for the construction of a viable German *modus vivendi*, and thus his entire journey stands under the "never-severed ties to German conditions, a tension which unifies the entire book and which we would call in today's jargon the

'Weimar Syndrome'" (Boerner 360). The crisis that drove Goethe from Germany, and his efforts at incorporating his self-transformation into Weimar culture on his return, form a shadow narrative to that of *IR* proper. His few literary works completed in Italy, such as *Iphigenie auf Tauris*, as well as those completed after his stay there, were intended to serve as models for German literature the way his construction of an Italian villa in Weimar was to serve as a model for German architecture and public/private space.

Landscapes of transit and of entry reveal above all the embeddedness of the traveler within the landscape of departure. Goethe signifies this revelation through the very mechanisms of geodesic location, longitude numbers. Consistent with this vision of a people imbricated in its landscape, Goethe insists on using spatial co-ordinates, whose largest dimensions are the lines of latitude crossed in his journey southward. He lends to each latitude line a significance which borders on the astrological. Goethe ties the 51st parallel, where he had been living in Weimar, to the continual diminishment of his talent and personality in the attrition of governmental and bureaucratic activity: "since my journey was in the nature of an escape from all the misfortunes I experience on the fifty-first parallel . . . I had hoped to enter a true Goshen on the forty-eighth" (*Italian Journey* 20; "meine Reise war eigentlich eine Flucht, vor allen den Unbilden, die ich unter dem ein und Fünfzigsten Grade erlitten, daß ich Hoffnung hatte unter dem Acht und Vierzigsten ein wahres Gosen zu betreten"; *Italienische Reise* 18-19). Goethe goes on to explain that his travels have taught him that simply going south does not necessarily alter the weather picture. Already in the first entry of 3 September 1786, the same day he steals away from his friends at 3:00 A.M.: "for otherwise I would never have gotten away" (13; "weil man mich sonst nicht fortgelassen hätte"; 9), by noon he finds himself at the same latitude as his home town of Frankfurt, and "I was happy to eat my midday meal on the fiftieth parallel again, under clear skies" (14; "und ich freute mich, wieder einmal bei klarem Himmel unter dem fünfzigsten Grade zu Mittag zu essen"; 9). On or about 5 September he enters yet another zone: "Today I am writing at the 49th parallel, which shows great promise" (14; "Heute schreibe ich unter dem neunundvierzigsten Grade. Er läßt sich gut an"; 11), he reports. On the sixth, he buys a fig in Munich, but finds its taste unworthy of the latitude, widening the complaint to a general breach of geographical *to prepon*: "in general the fruit is not especially good, considering that this is the forty-eighth parallel" (15; "das Obst überhaupt ist doch für den acht und vierzigsten Grad nicht besonders gut"; 12). He is in Munich and just before he reaches the 46th parallel, it seems to him that he has crossed the boundary of an entirely new world: "Written at the latitude of forty-five degrees and fifty minutes. In the cool of the evening, I went for a walk, and now I truly find myself in a new land, in a completely foreign environment. The people live a careless life of ease" (29;"Geschrieben unter dem fünf und vierzigsten Grade fünfzig Minuten. In der Abendkühle ging ich spazieren, und befinde mich nun wirklich in einem neuen Lande, in einer ganz fremden Umgebung, die Menschen leben ein nachlässiges Schlaraffenleben"; 29). This is written in Torbole, where Italian has become the dominant language.

Goethe has taken the latitudinal degrees designed to render continuous measurement, and converted them into a device for announcing sudden transformation. The comments of Italo Battafarano on this passage are of interest: "The point of departure [for Goethe's remarks] is his perception of the foreignness ['Fremdsein'] of the authorial 'I.' The traveler emphasizes decisively that the difference between Germany and Italy is not quantitative or gradual, but qualitative. Rather than speak of a 'different land' as one would normally do, Goethe speaks of a 'new land' that seems not just foreign, but 'completely foreign.' Goethe recognizes this foreign land not at the border between the two language areas, nor in the payment in foreign currency, nor in his transportation, nor in his eating, nor in his sleeping; rather, he recognizes it on an evening walk on the shores of Lake Garda" (131-32). Of course, only in a careless moment can a "careless life of ease" be perceived that seems to Goethe a remnant of an earlier stage of human existence, a foretaste of the "Homeric" conditions of Foligno. Hence his emphasis on the absolute alterity of Italy.

As he explains in his diary, Goethe measures parallels not so much to know the stations on the earth he is passing through, but the relation of earth to sky. Their conjunction now signal his finally reaching a land where the people live "under a happy sky of nature" (*Flight to Italy* 89; "unter einem glücklichen Himmel der Natur"; *Tagebuch* 285). After an extended stay in Venice, he takes up the latitudinal schema yet again, on the boat to Ferrara. Once again, climactic determinism shapes his remarks: "Now I have actually entered the forty-fifth degree of latitude and I repeat my old song: the inhabitants of this country could keep everything else if only, like Dido, I could lay thongs around enough of this climate to surround our dwellings at home. Really, it is a different world" (*Italian Journey* 84; "Ich bin nun in den fünf und vierzigsten Grad wirklich eingetreten, und wiederhole mein altes Lied: dem Landsbewohner wollt' ich alles lassen, wenn ich nur, wie Dido, so viel Klima mit Riemen umspannen könnte, um unsere Wohnungen damit einzufassen. Es ist denn doch ein ander Sein"; *Italienische Reise* 115).

Landscapes of nation

Goethe's basic idea is that the landscape of Italy exerts a special power to lift the veil from one's eyes, to change one's way of thinking, and to instill "la gaya scienza" of the "Frohblick," into German culture. Hence, the paradox of the *IR* as a landscape of nation: like Dante's journey *ad astram* which must begin with a downward motion, so the journey of German renewal must pass southward before returning to its Northern home. "It was in the paradise of the south that Goethe began to accustom himself to life in the German north" (Werner 38).

That contemporaries of Goethe discerned national themes in *IR* can be seen in the contrasting reactions of the historian Barthold Georg Niebuhr and the writer Ludwig Tieck. Niebuhr wrote from Rome to Friedrich Karl von Savigny on 16 February 1817 to complain that Goethe "considers an entire nation ['Nation'] and country ['Land'] simply as a delight designed for himself, and sees in all the world and nature nothing other than what belongs to the infinite decoration of his own pitiful life" ("so eine ganze Nation und ein ganzes Land bloß als eine Ergötzung für sich betrachtet, in der ganzen Welt und Natur nichts siehet, als was zu einer unendlichen Dekoration des erbärmlichen Lebens gehört"; qtd. in Bode 9). Niebuhr's letter has often been cited as a severe critique of the *IR* and as indicative of its relatively poor reception; however, I wish to point out here that in his criticism Niebuhr in fact correctly identifies the process by which Italy as nation and country has been sublimated into the Goethean gaze, whose vision of Italy in turn became central to his conception of what Germany should be.

Goethe's preoccupation with latitude may derive at least in part from the inadequacy of national boundaries in his period to define location. In Ekkehart Krippendorf's analysis, Goethe had numerous models of nation available to him, from the *Kleinstaat* of Weimar through the nation-states of France and Prussia, to the federative Holy Roman Empire. His choice was for the *Kulturnation*: "Goethe saw political world-citizenship as a serious and desirable alternative to the nation-state and saw the culture-nations as organic parts of humanity, while the nation-state represented a fatal wrong turn" (217). Hence the erasure of political boundaries in *IR*. The fact that Goethe had left the kingdom of Bohemia (where he was vacationing at Karlsbad) and entered the state of Bavaria was unimportant to him, and he does not mention the fact. Similarly, he never writes, "I have arrived in Italy." No moment of border crossing is ever portrayed in the *IR*, although such moments must surely have occurred. The fact is that German is spoken in München, as it is in Bozen/Bolzano, and hence these places are for Goethe stages on the way towards the new, which is defined in terms of language, culture, art, latitude, and landscape.

What Goethe saw in Italy between 1786 and 1788 went beyond expanding the boundaries of his mental map of Europe; it influenced his "way of seeing," to use John Berger's simple yet powerful phrase. The way of seeing Goethe developed in Italy, Mikhail Bakhtin notes, "searches for and finds primarily the visible movement of historical time, which is inseparable from the natural setting . . . and the entire totality of objects created by man, which are essentially connected to this natural setting" (32). Upon his return to Naples from Sicily, Goethe wrote to his friend, Johann Gottfried Herder, in Weimar: "when I return home you shall judge how I have seen" (*Italian Journey* 255; "wenn ich wiederkomme, sollt Ihr beurteilen, wie ich gesehen habe"; *Italienische Reise* 322). Goethe considers the judgment by others of his own vision as integral to his project. His own self-fashioning was meant to have repercussions on the society he returned to, that of the Weimar court and its resident intellectuals. The author's inner vision of Italian landscape was to be publicized for a German audience. The element of reflexivity in every vision he reports derives from this translation effort.

Ways of seeing, ways of narrating

In 1788, near the end of his stay in Rome, Goethe notes one of these private epiphanies, a change in his perception of the Italian landscape: "Only now are the trees, the rocks, indeed Rome itself becoming dear to me; heretofore I have always just felt that they were foreign. On the other hand, I was pleased with unimportant objects that were similar to what I saw in my youth. Now I must begin to feel at home here also, and yet it will never be as close to me as those first things in my life. In this connection, I have had various thoughts concerning art and imitation" (*Italian Journey* 278; "Jetzt fangen erst die Bäume, die Felsen, ja Rom selbst an mir lieb zu werden; bisher hab ich sie immer nur als fremd gefühlt; dagegen freuten mich geringe Gegenstände, die mit denen Ähnlichkeit hatten, die ich in der Jugend sah. Nun muß ich auch erst hier zu Hause werden und doch kann ichs nie so innig sein als mit jenen ersten Gegenständen des Lebens. Ich habe verschiedenes bezüglich auf Kunst und Nachahmung bei dieser Gelegenheit gedacht"; *Italienische Reise* 351).

Chloe Chard has noted that eighteenth-century travelers characteristically "construct oppositions between the foreign and the familiar even when commenting on relatively trivial aspects of the topography of foreignness" (42), but this passage moves beyond opposition to an involved dialectic of familiarity and strangeness. Firstly, the diataxis moves us from mimesis of nature (the rocks, the trees) to a diegesis not of Roman culture, but of Goethe's own *Bildung*. Secondly, one notes the zeugma in linking rocks and trees to Rome. One would expect, of course, in the place of these elements of landscape, to encounter urban phenomena such as buildings and streets, and customs such as the festivals Goethe describes elsewhere. One would expect to find people, and the passage thus demonstrates that in Italy, "the people are for [Goethe] a part of nature, not of history" (Althaus 150-51). Thirdly, Goethe implies here the need to publicize his private epiphany in the form of writing on art. He considers the judgment by others of his own vision as integral to his project. His own self-fashioning was meant to have repercussions on the society he returned to, that of the Weimar court and its resident intellectuals.

The price of changing his vision, Goethe, never tires of repeating, is high, and the process lengthy:

> If one considers that for centuries the finest architecture flourished here, that the artistic ideas of first-rate minds took visible form and rose up on mighty substructures remaining from ancient times, then one can understand how charmed the eye and mind must be when, in every kind of light, one sees these multiple horizontal and thousands of vertical lines, interrupted and adorned, like a mute music, and how, not without pain, everything petty and narrow-minded in us is dislodged and driven out. Especially indescribable is the richness of the moonlight scenes, where individual enjoyable—perhaps they should be called intrusive—details recede, and there are only the great masses of light and shadow, which fill the eye with hugely graceful, symmetrically harmonious, colossal shapes. (326)

> Wenn man bedenkt, daß Jahrhunderte hier im höchsten Sinne architektonisch gewaltet, daß, auf übrig gebliebenen mächtigen Substruktionen, die künstlerischen Gedanken vorzüglicher Geister sich hervorgehoben und den Augen dargestellt: so wird man begreifen wie sich Geist und Art entzücken müssen, wenn man unter jeder Beleuchtung diese vielfachen horizontalen

> und tausend vertikalen Linien unterbrochen und geschmückt wie eine stumme Musik mit den Augen auffaßt, und wie alles, was klein und beschränkt in uns ist, nicht ohne Schmerz, erregt und ausgetrieben wird. (491)

In a diary passage, Goethe notes that part of this "natural history" of Italian architecture is the public use of buildings erected by the upper classes:

> Though I don't know how things look inside their palazzi. The forecourts, arcades, etc. are all filthy with refuse, and that's quite natural, you just have to see it from the people's angle. They feel they have prior rights. The rich man can be rich and build palaces, the *nobile* is free to rule, but if he puts up an arcade or a forecourt, then the people use it. . . . Anybody who doesn't want that mustn't play the great lord; that is to say, he mustn't behave as if a part of his dwelling-place were a public space, he has to shut his door and then everything's clear. With public buildings, the people won't be done out of their due. And that's the way it is everywhere in Italy. One more observation that isn't easy to make. (*The Flight to Italy* 36)

> Doch weis ich nicht wie es im Innern ihrer Palazzi aussieht. Die Vorhöfe, Säulengänge sind alle mit Unrath besudelt und das ist ganz natürlich, man muß nur wieder vom Volck herauf steigen. Das Volck fühlt sich immer vor. Der Reiche kann reich seyn, Palläste bauen, der Nobile darf regiren, aber wenn er einen Säulengang, einen Vorhof anlegt, so bedient sich das Volck dessen zu seine Bedürfniß. . . . Will einer das nicht haben; so muß er nicht den Grosen Herren spielen; das heist, er muß nicht thun als wenn ein Theil seiner Wohnung dem Publiko zugehöre, er muß seine Thüre zumachen und dann ists gut. An öffentlichen Gebäuden läßt sich da Volck sein Recht nicht nehmen. Und so gehts durch ganz Italien. Noch eine Betrachtung die man nicht leicht macht. (*Tagebuch* 65-66)

The two observations in these quotes seem to lie on the same order of difficulty. Both are part of the process of *Bildung*, in that they implicitly compare German with Italian reality. We must ascribe the difficulty of arriving at this conclusion to the difference between Italian and German "public spheres." Goethe counters here a number of "enlightened" or "rationalistic" travel narratives that point out the uncleanliness and laziness of the Italians with a political interpretation of the relation between grandiose architecture and its appropriation by the people. The difference may be captured by turning the description of an architectural image into a narrative of its use. The story Goethe tells here might stand for his conception of the best relation to be hoped for between ruler and ruled. Not democratic or egalitarian, but with the people retaining important powers and the nobility important obligations. Of course, such a vision hardly tends to the ideal; the power of the people is expressed in their right to dirty the atria and colonnades of semipublic buildings, hardly an uplifting activity. It is also expressed in the thirty-eight official holidays, such as Carnival, reduced from more than one hundred a few years before Goethe's visit. Goethe's striking of the passage from *IR* derives perhaps from his hardened position against populism following the French Revolution and Napoleon.

The story of Goethe's return to Germany, on the other hand, is told not in *IR*, but in one of the many apparatuses attached to the *Metamorphose der Pflanzen*, a book inspired by the trip to Italy. Complementary to the gradual coming into clarity of the trip southward, the return northward was a dissolution of form: "Out of polymorphic Italy I was exiled to formless Germany, to trade a clear sky for a cloudy one. . . . no one was understanding my language" ("Aus Italien dem Formreichen war ich in das gestaltlose Deutschland zurückgewiesen, heiteren Himmel mit einem düsteren zu vertauschen. . . . niemand verstand meine Sprache" ["Schicksal" 102]). Everything is reversed, including the terms of at-homeness and strangeness indicated by the comprehension of language. It is as if Goethe had returned to Germany speaking Italian. In Italy, Goethe learned "to differentiate between what is akin and what is foreign to me" (*Italian Journey* 277; "unterscheiden, was mir eigen und was mir fremd ist"; *Italienische Reise* 350). *IR* offers many examples of such sensuous material—art, architecture, volcanoes, *carnevale*, and so on—but as Bakhtin reminds us, all these can be subsumed under the category of *Landschaft*. The assimilation is never complete: Goethe does not become Italian. Rather, Italy allows him to recognize why first objects ("erste Gegenstände") provide the steadiest landmarks for self-orientation and impede his full transformation. Halfway through his stay he writes, "I feel easier in my mind, and am almost not the same person I was a year ago" (306; "Ich fühle mir einen leichtern Sinn und bin fast ein andrer Mensch als vorm Jahr"; 383). Goethe's approach to the temple of Minerva in Assissi shows how narrative and description combine to develop the character of the narrator. He leaves the coach, wishing "to wander on foot through what seemed through what seemed to me a very isolated world" (96; "durch die für mich so einsame Welt eine Fußwanderung anzustellen"; 133). He enjoys his view of the symmetrical architecture and the temple's location, which Palladio had left out of his drawing. He lets the view of the temple transform him: "Contemplation of this work awakens feelings in me that I cannot put into words, but they will bear lasting fruit" (97-98; "Was sich durch die Beschauung dieses Werks in mir entwickelt, ist nicht auszusprechen, und wird ewige Früchte bringen"; 137). This is the first of many "hidden" narratives Goethe keeps secret in a game of *trompe l'oeil* with his readers. He then chooses to narrate a brief story, in which reality intervenes to interrupt contemplation: "It was the most beautiful evening, and I was walking downhill on the Roman road, wonderfully calm in spirit, when behind me I heard rough, vehement voices raised in a quarrel" (98; "Ich ging am schönsten Abend, die römische Straße bergab, im Gemüt zum schönsten beruhiget, als ich hinter mir, rauhe heftige Stimmen vernahm"; 137). The owners of these voices surround Goethe and question his presence at the Minerva temple. Goethe eventually calms their suspicions that he is a smuggler, but the incident gives the rest of his walking tour a peculiar, dark twist. In moving from description to narration, Goethe also places himself more in the foreground, makes himself stand out dramatically from his surroundings, and thus heightens the contrast between his foreignness and the Italy that will provide him the contrasts against which he can mold himself. From Greek architecture (which shocked Goethe with its raw massiveness when he first beheld it at Paestum), life falls back into the prehistoric, as Goethe notes that the residences of this area highly resemble caves. Their inhabitants make no efforts to improve their residences, nor even to prepare for the winter. It is as if the "rough voices" had condensed and concretized themselves into rough surroundings. Narrative has changed back again into description. Goethe dramatizes further the atavistic landscape by letting it affect his ability to write in the most concrete way: "Here I am in Foligno, in a truly Homeric domestic situation, where everyone assembles around a fire burning on the earthen floor of a great hall and, amidst shouting and clamor, dines at a long able, as in paintings of the marriage at Cana. Under the circumstances I would not have thought of inkwells, but somebody has had one brought in, and so I seize the opportunity to write this. But the page reveals how cold and uncomfortable my writing table is" (99; "Hier in Foligno, in einer völlig homerischen Haushaltung, wo alles um ein auf der Erde brennendes Feuer, in einer großen Halle versammelt ist, schreit und lärmt, am langen Tische speist, wie die Hochzeit von Cana gemalt wird, ergreife ich die Gelegenheit dieses zu schreiben, da einer ein Dintenfaß holen läßt, woran ich unter solchen Umständen nicht gedacht hätte. Aber man sieht auch diesem Blatt die Kälte und die Unbequemlichkeit meines Schreibtisches an"; 139). And the first words of the next entry are simply "Here I sit in another cave" (100; "Wieder in einer Höhle sitzend"; 140). Narrative, classical and Biblical allusions, and self-reflexive description of the conditions he writes in, all contribute to Goethe's intercalation into a landscape.

In a similar fashion, Goethe moves between narrative and description in the pages he devotes to the Santa Rosalia church built into the cliffs above Palermo. After describing the technical apparatus used for carrying away the water from what is essentially a cave, he approaches the altar with its statue of Rosalia: "By the gleam of some subdued lamps I caught sight of a beautiful woman. She lay as though in a kind of rapture, her eyes half closed, her head carelessly laid on her right hand, which was adorned with many rings. I could not get enough of looking at this image; it seemed to me to have quite special charms" (*Italian Journey* 193-94; "Ein schönes Frauenzimmer erblickt' ich bei dem Schein einiger stillen Lampen. Sie lag wie in einer Art von Entzückung, die Augen halb geschlossen, den Kopf nachlässig auf die rechte Hand gelegt, die mit vielen Ringen geschmückt war. Ich konnte das Bild nicht genug betrachten; es schien mir ganz besondere Reize zu haben"; *Italienische Reise* 296). He goes on to demonstrate the truth of his statement that he could not get enough of Rosalia's image. The description of his trance before the image forms perhaps the greatest time-dilation in the whole *IR*. Goethe struggles against the narrative impedance caused by the fact that he is describing absolute stillness and trance. The more nothing is done, the more words are needed to describe the *satori* that the meditator experiences. The monks enter and sing vespers, he listens awhile, and then approaches the altar again. He kneels, and tries to see "the beautiful image of the saint still more clearly. I yielded completely to the charming illusion of the figure and the place. Now the priest's song died away in the cavern, the water trickled down and gathered in the container next to the altar, the overhanging rocks of the outer court, which is the actual nave of the church, enclosed the scene still more tightly" (194; "das schöne Bild der Heiligen noch deutlicher gewahr zu werden. Ich überließ mich ganz der reizenden Illusion der Gestalt und des Ortes. Der Gesang der Geistlichen verklang nun in der Höhle, das Wasser rieselte in das Behältnis gleich neben dem Altare zusammen, die überhangenden Felsen des Vorhofs, des eigentlichen Schiffs der Kirche, schlossen die Szene noch mehr ein"; 296). His description moves outward, from the momentary and transitory, such as the musical setting, to the essential, the telluric setting of the church. The totality of the aesthetico-religious experience, merging art, architecture, and music within a palpable topographic framework, and uniting present and past, is the equivalent of the *Urpflanze* that Goethe sought in Italy. There is a reason for the existence of this church, a reason allied with the earth and its secrets, yet found only in the experience of the moment, an instant capable of expansion into seeming infinity. Goethe ends his description with the notation that he arrived back in Palermo only "late at night" (194; "erst in später Nacht"; 296). The striking ellipsis between the few sentences of narrative and the end of the story encompasses this infinity.

On the whole, then, *IR* tends towards engaged narrative. Missing from it is any extensive, distanced, disengaged description of Italy as a whole culture or landscape. (The closest we get, as I will show below, is Italy as a climatological unit.) Herbert Lehmann has noted *IR*'s avoidance of totalizing description: "*IR* breathes landscape in almost every line, but if one looks for quotes to confirm this overall impression, one finds a large number of exact observations, characterized for the most part by a lengthy, scientific description of landscape" (176). In other words, Goethe himself claims, with some exaggeration, as I have shown, that he takes up a non-narrative approach deliberately to the landscape: "If, instead of losing oneself in fantasy here, one accepts the region as reality, just as it lies there, then it is still the definitive scene of action, which calls for the greatest deeds; and therefore up till now I have always made use of my interest in geology and topography to suppress my imagination and sentiment, and to preserve a free, clear view of the locality for myself" (101; "Wenn man hier nicht phantastisch verfährt, sondern die Gegend real nimmt, wie sie daliegt, so ist sie doch immer der entscheidende Schauplatz, der die größten Taten bedingt, und so habe ich immer bisher den geologischen und landschaftlichen Blick benutzt, um Einbildungskraft und Empfindung zu unterdrücken und mir ein freies klares Anschauen der Lokalität zu erhalten"; 122).

Exceedingly familiar with landscape painting—he mentions numerous landscape painters in his narrative, from Claude Lorrain to Aert van der Neer—Goethe recognized intuitively what the genre, a relatively new one in the history of painting, meant: the subordination of topographic and cultural features to a single, coherent, dominating gaze. Goethe interprets the social relations of the Italian villa through its organization under the dominant gaze of the *signore*: "We saw a splendid, although not unexpected, view from the windows of the villa of Prince Aldobrandini. . . . As can be imagined, the mansion was designed in such a way that the magnificence of the hills and plain could be surveyed from it at a glance" (*Italian Journey* 327; "Eine herrliche, obgleich nicht unerwartete Aussicht ward uns aus den Fenstern der Villa des Fürsten Aldobrandini. . . . Es läßt sich denken, daß man das Schloß dergestalt angelegt hat, die Herrlichkeit der Hügel und des flachen Landes mit einem Blick übersehen zu können"; *Italienische Reise* 492). Why "not unexpected"? Perhaps Goethe is thinking here of the requirements of the landscape genre as practiced in his time. Here, landscape equals social system. Goethe intuitively grasps the "analogy between the rural estate and the state of the realm which is so frequently made in poetry and in the country house picture," an analogy that "reinforces the ideology of landscape in the service of absolutism. A well-managed country house and its lands form a self-sufficient world, a microcosm of the mercantilist state . . . But [that world's] harmony rests ultimately on its subordination to the care and authority of one all-powerful lord, as the harmony of its landscape depends on the lordship of the eye" (Cosgrove 196). Goethe was not an apologist for absolutism; however, as I mentioned previously, he did consider common Germans to be in need of guardianship by a powerful authority until the time when *Bildung*, cultural awareness, could elevate them to the necessary state of political awareness where they could determine their own destiny. His description of the aesthetically placed mansion parallels both the seemingly disinterested world of landscape painting, and a political view of order and control emanating from a single, elevated perspective. Furthermore, recalling the Dido quote, Goethe implies here that control extends only to the end of one's visual field. It provides a diametric contrast with the relatively "flat" and unnavigable worlds of Gogol and Turgenev. In other cases, however, Goethe resists those narratives that do not include him. For example, he rejects the efforts of his Paleman guide to vivify the landscape with a report of Hannibal's activities there:

> The most beautiful spring weather and a burgeoning fertility spread a feeling of refreshing peace over the whole valley, but it was spoiled for me by the pedantry of our inept guide. He related in detail how Hannibal had once fought a battle here, and told of the other great warlike deeds that had occurred in this place. I crossly rebuked him for so wretchedly evoking these departed spirits. I said it was bad enough that the crops had to be trampled down from time to time, if not always by elephants, then by horses and men; but at least my imagination should not be startled out of its peaceful reverie by these tales of tumult. He was quite astonished that I should spurn classical memories in a place like this, and of course I was unable to make him understand how such mingling of past and present affected me. (*Italian Journey* 189)

> Die schönste Frühlingswitterung und eine hervorquellende Fruchtbarkeit verbreitete das Gefühl eines belebenden Friedens über das ganze Tal, welches mir der ungeschickte Führer durch seine Gelehrsamkeit verkümmerte, umständlich erzählend, wie Hannibal hier vormals eine Schlacht geliefert und was für ungeheure Kriegstaten an dieser Stelle geschehen. Unfreundlich verwies ich ihm das fatale Hervorrufen solcher abgeschiedenen Gespenster. Es sei schlimm genug, meinte ich, daß von Zeit zu Zeit die Saaten, wo nicht immer von Elephanten doch von Pferden und Menschen zerstampft werden müßten. Man solle wenigstens die Einbildungskraft nicht mit solchem Nachgetümmel aus ihrem freidlichen Traume aufschrecken. Er ver-

> wunderte sich sehr, daß ich das klassische Andenken an so einer Stelle verschmähte und ich konnte ihm freilich nicht deutlich machen, wie mir bei einer solchen Vermischung des Vergangenen und des Gegenwärtigen zu Mute sei. (*Italienische Reise* 290)

Goethe refuses to hear this story, wishing only to create his own diegeses from the mimesis of landscape. Goethe himself gives at least two reasons for banishing the story of Hannibal from the scene, both involving the principle of *to prepon*: a feeling of peace should not be disturbed by stories of war; and present and past should not be jumbled together. This incident should temper the prevailing view of the *IR* as a classicizing document. Here we see resistance to history's claims on the Sicilian landscape. Goethe refuses to hear this story, refuses to mix present with past, geography with history, and turns instead to assembling pebbles which he will be able to shape into an ostensibly more useful diegesis, that of the landscape itself and of how it came to be: "It seemed still odder to our escort when I looked in all the shallows, many of which the river leaves quite dry, for little stones, and took along specimens of the various kinds. Again I was unable to explain to him that the quickest way to understand a mountainous region is to inspect the minerals swept down by the brooks, and that here too the task was to use rubble in order to obtain an idea of those earthly antiquities, the eternally classical mountains" (189-90; "Noch wunderlicher erschien ich diesem Begleiter, als ich auf allen seichten Stellen, deren der Fluß gar viele trocken läßt, nach Steinchen suchte und die verschiedenen Arten derselben mit mir forttrug. Ich konnte ihm abermals nicht erklären, daß man sich von einer gebirgigen Gegend nicht schneller einen Begriff machen kann, als wenn man die Gesteinarten untersucht die in den Bächen herabgeschoben werden und daß hier auch die Aufgabe sei, durch Trümmer sich eine Vorstellung von jenen ewig klassischen Höhen des Erdaltertums zu verschaffen"; 290).

Goethe, no less than his troublesome guide, links present and past in his own narrative. He uses the adjective "klassisch" again, but with its meaning now in the register of natural rather than cultural periodization, thus implicitly establishing an opposition between historical and natural historical narration, between narratives of historical actors in a natural setting versus narratives of the actions of landscape itself. Bakhtin reads the incident as pointing out, first, Goethe's dislike of the "estranged past" in favor of understanding the "necessary place of this past in the unbroken line of historical development," and second, Goethe's principle that "the past must be creative" (33-34). To this we should add, however, that the narrative Goethe constructs from pebbles belongs entirely to him. He does not even share it with his readers, keeping it as another hidden narrative whose invisibility increases the narrator's power, just as the landlord's gaze rests as much in its power to hide as to reveal. The aesthetic pleasure of the controlling gaze from the Aldobrini house lies in part in one's ability to see without being seen, a principle articulated most notably by Jay Appleton in *The Experience of Landscape*. Contrary to expectations, he found Sicily to be a place of transience and violence that disrupted the aesthetic of vision and landscape I have been describing above.

Geodetic differences

Goethe left out deliberately in *IR* more direct political statements about this "new land" recorded in his diary, such as the following: "In a country where people enjoy the daytime, but especially look forward to the evening, the point when it becomes night is especially significant. When the day's work ends. When people have to start their stroll or come back home. When night falls, fathers want to have their daughters back indoors and so on, night concludes evening and ends the day. And what a day means is something that we Cimerians in our eternal mists and gloom scarcely know, it's all one to us whether it is day or night, for when can we ever enjoy ourselves in the open air?" (*The Flight to Italy* 41; "In einem Lande wo man des Tags genießt, besonders aber sich des Abends freut, ist es höchst bedeutend wenn es Nacht wird, Wann die Arbeit des Tages aufhöre? Wann der Spaziergänger ausgehen und zurückkommen muß. Mit einbrechender Nacht will der Vater seine Tochter wider zu Hause haben pp. die Nacht schließt den Abend und macht dem Tag ein Ende. Und was ein Tag sey wissen wir Cimmerier im ewigen Nebel und Trübe kaum, uns ist's einerley ob's Tag oder Nacht ist, denn welcher Stunde können wir uns unter FREYEM HIMMEL freuen?"; *Tagebuch* 286). Goethe plays here with the double meaning of "free sky" (an ambiguity lost in the published translation, which prefers "open air"): without clouds, but also metonymically indicating political and social freedom. Klaus Kiefer comments that this passage reflects the "concrete utopia": Italy's "special feature, the climatic advantage, is made into a mytheme of general freedom through the anthropomorphic attribute 'happy'" (343).

Counter-narratives, counter-mappings

Italienische Reise can be considered a counternarrative to selected Italian travel books that came before it. Previous scholarship has provided a thorough listing of these narratives (see Schudt; Klenze). Goethe takes aim at two main groups of texts on Italy, which Camillo von Klenze calls the Rationalistic and the Romantic. "Rationalistic" travel narratives tended to give small place to medieval and early Renaissance art, and to highlight the backwardness of Italian economic, social, and political institutions in comparison especially with those of England. D.J.J. Volkmann's *Historisch-kritische Nachrichten von Italien* of 1770 served both as Goethe's main guidebook in Italy, and also as a main target of his counter-narrative. The passage on the presence of garbage in the atria and colonnades of great houses cited above furnishes a good example of Goethe's ability to insert behavior into his narrative that a rationalist would simply criticize as deficient or uncivilized. The so-called "Romantic" travel narratives—published after Goethe's stay in Italy, but before the publication of the *IR*, are characterized by their confrontation with the French occupation of Italy, and by their subjective and often lyrical attitude. Despite its high degree of self-reflexivity, the *IR* emphasizes exact observation rather than inspiration. As Hans-Georg Werner has noted, Goethe's sharp demarcation of Italy as a new, happy, public country, as Germany's Other, would not have corresponded to the preconceptions of readers of the first published version of *IR* in 1816. The Napoleonic wars had impressed them instead with the overall unity of Europe, and with the invisibility and permeability of the borders between nations. Travelers to Italy began seeing not the antique, but the modern, the political process of democratization and nation building. Goethe recognizes this fact as well in other writings, as noted above, which makes the publication of *IR* a deliberate act of nostalgia, of subtle resistance both to the "New European order" designed by the Congress of Vienna, and to German nationalism that rose in response to Napoleon's invasion and the dissolving of the Holy Roman Empire. His approach to Italy has been contested from the first appearance of *IR*. According to Gretchen Hachmeister, "the literary Italienbilder of the 1820s manifest a vast range of response to Goethe's experience of Italy, including incomprehension, avoidance, and rejection" (177).

In a long, detailed, and insightful review of *IR* at its first appearance, Ludwig Tieck was moved to comment on the relationship between time, space, and nationality in the work:

> [Goethe] forgets . . . that our pure longing for what has been lost, where nothing of the present can disturb us, transfigures these relics and fragments and transports them into the pure domain of art. This domain, however, has never been such as to permit us to devalue our customs, fatherland, and religion on its behalf. . . . I, too, have visited St. Peter's and antiquities, and they only increased my admiration for the Strassburg cathedral. Only after learning Raffael by heart, did I fully understand the worth and attractiveness of Old German art. . . . I love the Italians and their ways, but I became a German only in Italy. There is no such thing as a poet without a fatherland!

> [Goethe] vergißt . . . daß unsere reine Sehnsucht nach dem Untergegangenen, wo keine Gegenwart uns mehr stören kann, diese Reliquien und Fragmente verklärt und in jene reine Region der Kunst hinüberzieht. Diese ist aber auch niemals so auf Erden gewesen, daß wir unsere Sitte, Vaterland und Religion deshalb geringschätzen dürften. . . . Ich hatte auch die Antike gesehen, Sankt Peter, und konnte den Straßburger Münster nur um so mehr bewundern. Nach dem auswendig gelernten Raffael verstand ich erst die Lieblichkeit und Würde altdeutscher Kunst—und dies wäre Oberflächlichkeit, Einseitigkeit etc. in mir gewesen? Ich liebe die Italiener und ihr leichtes Wesen, bin aber in Italien erst recht zum Deutschen geworden. . . . Ohne Vaterland kein Dichter! (qtd. in Bode 668)

Tieck uses the occasion of Goethe's *IR* to restate the vision of German art he and Wackenroder had brought onto the public stage in works such as the *Reveries of an Art-Loving Monk* (1797; "*Herzensergießungen eines Kunstliebenden Klosterbruders*"), a book generally considered to be the beginning of the Romantic movement in Germany. Tieck also begins a tradition of providing counter-narratives to *IR*—he discovers the wonders of the Strassburg cathedral after his trip to Italy, reversing the chronological order of Goethe's acquaintance with monuments. Tieck will not admit that his comment that it was Italy that had made him a true German applies to Goethe in *IR* as well, in the very process of *Bildung*. Given that Niebuhr chides Goethe for neglecting the Italians, while Tieck censures him for forgetting the Germans, we may suspect that a more accurate reception of *IR* lies in a dialectical reading in which a vision of Italy always implies a corresponding one of Germany, and vice versa.

This tension between pre- and post-Napoleonic ways of seeing national landscapes dominates the approach Italian novelist Alessandro Bárbero has taken to turning Goethe's traveler's viewpoint inside out. His diary novel, *Beautiful Life and Foreign Wars of Mr. Pyle, Gentleman* (1995; *Bella vita e guerre altrui de Mr. Pyle, gentiluomo*), can be said to fulfill Goethe's idea of *Weltliteratur*, bringing together as it does the viewpoints of Italy, Germany, and the United States. Mr. Pyle is a US-American appointed ambassador to the King of Prussia as the Napoleonic wars are about to commence. Pyle travels through all of Prussian territory, from Brandenburg to Königsberg to Warsaw. Pyle has little to say about art or landscape, other than the atrocious East Prussian roads and the hideous poverty in German villages. Instead, he focuses his memoirs on his conversations with princes and ministers, largely on the question of what to do about Napoleon. Pyle ends his journey in Weimar, where he dines with and interviews Duke August and his minister, Goethe. The last fifty pages of the novel show Pyle observing the outbreak of the war which brings a disastrous defeat for the Germans. The choice of a US-American as observer also implies that all of Europe will soon be eclipsed under the economic and military power of the United States.

Pyle's description of Weimar seems calculated to affirm Goethe's aversion that drove the poet south. Unlike the Italian cities described by Goethe, Weimar offers no sheltering architecture to support the great artistic endeavors of its privileged inhabitants: "Weimar . . . is a poor dump of houses stained by smoke and falling down, and of roads so poorly maintained that at any moment a wagon full of bread or beer for the soldiers, a caisson full of munitions for the artillery, an ambulance heading for the field hospital get stuck in the mud, obstructing the passage of other vehicles and, where the houses are close together, of pedestrians as well" ("Weimar . . . è un buco miserabile, dalle case affumicate e cadenti, dalle strade così malandate che ad ogni momento un carro carico di pane o di birra per i soldati, un cassone di munizioni dell'artigliera, un'ambulanza diretta al lazzaretto rimanon presi nel fango, ostacolando il passaggio alle carroze e, dove le case sono più vicine l'una all'altra, perfino ai pedoni"; 496). Similarly, Pyle's first impression of Goethe is of a cold, unfriendly man barely capable of maintaining the formalities of conversation at table (Goethe is upset because he has had to dislodge from the palace to make room for a visitor of higher rank). But as the gentlemen relax over wine after dinner, Goethe reappears and begins to discourse on the new era promised by Napoleon. The contrast Goethe makes between a formless, amoral French *peuple* and the great leader who has made them into something reverses the obsession found in *IR* with a leaderless Italian people so fully formed that they have become a fixture of landscape. Under Napoleon, the French are experiencing *Bildung*:

> Anyone who sees the French up close can only feel repugnance. They are all educated, but the last crumb of moral sense has expired in them. . . . But see how a great hero has emerged from the disturbances of the great

> Revolution, one capable of molding out of nothing a people previously without form, and of giving new life to the spent material. And his marvelous genius throws his shadow even over our poor German people. . . . I didn't used to believe in the possibility of a universal empire, yet this is realizing itself before our eyes. He already is the master of France and Italy, of Naples and Holland. Spain and Switzerland would not budge without his permission, the German princes have submitted to him.
>
> Chiunque veda da vicino i francesi non può non provare un'autentica ripugnanza. Sono tutti istruiti, ma l'ultima favilla di senso morale in loro è spenta. . . . Ma ecco che dagli sconvolgimenti della grande Rivoluzione è emerso un eroe capace di plasmare dal nulla quel popolo ormai informe, di ridare la vita alla materia spenta, e il suo genio merviglioso a getta la sua ombra anche sul nostro povero popolo tedesco. . . . Non credevo alla possibilità di un impero universale, ed esso si sta realizzando sotto i nostri occhi: egli è già padrone della Francia e dell'Italia, di Napoli e dell'Olanda, la Spagna e la Svizzera non muoverebbero un dito senza il suo permesso, i principi tedeschi si sono sottomessi. (521)

Pyle encounters Goethe once more, by chance, as he saves Goethe's housekeeper and lover, Christiane Vulpius, from harassment and mugging, and she subsequently leads him to the Frauenplan house and Goethe (he had taken Vulpius into his house shortly after his return from Italy; he married her soon after Weimar fell to the French). Pyle finds Goethe strangely calm about the approaching French troops which may well end up extinguishing Weimar as an independent dukedom. Pyle then accompanies the Duke's troops as they are defeated by the French. He is captured by the French commander and then released upon recognizance of his diplomatic status.

Like Goethe's *IR*, Bárbero's narrative defies easy generic categorization. The fictional character Pyle, like other invented characters in the narrative, seems to be there for the sole purpose of allowing the reader to see and hear historical personages such as the Duke of Weimar and Goethe. By making Pyle a US-American, Bárbero draws implicitly analogies between the Napoleonic New European Order and the twentieth-century version of leadership in the United States. It is no accident that Bárbero is a historian by profession with several published works on the Middle Ages and Charlemagne. The resulting "faction" was nominated for several literary prizes, and won the prestigious Strega. Bárbero's account shows an intimate acquaintance with Goethe's writings and with accounts of him by others. Goethe realized more quickly than his contemporaries the meaning of the French Revolution and of Napoleon, and then let himself get caught up in those consequences to a lesser degree than they did. It is true that in *IR* he passes over those changes deliberately; indeed, its publication in 1816 represented a kind of literary restoration parallel to the Congress of Vienna. By recreating himself as a fictional traveler, Goethe bracketed the time interval between the experience of the moment and its recollection in later life, from a very different perspective. He bracketed therewith also the political and social changes of the three decades that intervened between experience and initial recording, and publication of the completed work.

Bárbero's narrative participates in the extensive tradition of exchange of fictions between Italian and German authors, while rising above the clichés of most products, just as Goethe's text transcends the genre of the travel narrative that had helped him navigate through Italy (for an account of this tradition, see Beller). As a historian, Bárbero's attempt at relocating the emergence of the nation-state from Napoleonic destruction seems aimed at recovering for his reader a delicate balance. Other post-war narratives of Italian travels, such as Rolf Dieter Brinkmann's *Rom, Blicke* (1979; hereafter, *Rom*), on the other hand, deconstruct the idea of nation. Brinkmann focuses neither on the differences between Germans, Italians, and Austrians, nor on the wonders of Roman architecture and city life. Even climatic observations—other than of smog and pollution—rarely appear in the course of more than 400 pages recording his months in Rome. Brinkmann instead places emphasis on the disappearance of all individuality and of any aesthetic sensibility under the triumph of international capitalism and its *Massenkultur*. Like Goethe, Brinkmann escaped from an uncomfortable situation in Germany—including, in this case, both his married life and the hangover of the 1960s student movement (which he detested)—by spending several months in Rome as a fellow of the German government, parallel to Goethe's generous support by the Duke of Weimar. Like *IR*, *Rom* is constructed out of a variety of materials: letters and postcards, diary entries, photographs, maps, restaurant bills, and excerpts from literature. Unlike Goethe, Brinkmann did not seek to hide the heterogeneity of this material. On the contrary, he made the text of *Rom* deliberately to look like a scrapbook as part of a technique called "cut-up." "Cut-up," borrowed by Brinkmann from William Burroughs, seeks to juxtapose radically different texts, thus emptying them of fixed meaning and allowing the language or image to speak "for itself." Letters alternate with diary entries, which in turn contain descriptions, theory, political observations, accounts of money spent, and stream-of-consciousness writing. In the opinion of Hermann Schlösser, "what makes Brinkmann's Rome book interesting as travel literature is above all the intensity of perception in the cut-up views. The meandering, gliding prose that corresponds to this intensity accompanies the unfolding of the journey" (98). Implicitly, Brinkmann's text rejects any attempt at the "intercalation of the writing self," although Brinkmann is always the focus of our attention. "The category of the sudden, the lightning-fast blending of disparate elements into a new insight that furthers our field of perception, anchors the writing so solidly, that one may say that it characterizes Brinkmann's style" (Adam 230). The anchor of our reading experience of *Rom* is not an intercalated self, but rather the very principle and feeling of motion and in-betweenness that precludes a sense of self, what Hans Adler had called the "subjectively experienced lucidity of the here and now" (99). A subjectivity concerned only with the present, occupied fully by aesthetic feelings, is of course not a product of *Bildung* and falls outside classical ideas of the subject, approaching instead the deconstructive psychoanalytic views of Jacques Lacan. In Sibylle Späth's words, "the questioning of the modern subject that plays a central role in Brinkmann's travel diary is a long way from the lofty tone of self-conscious subjectivity of the late-eighteenth century" (103). The lack of a central consciousness in favor of, in Adler's reading, an *aisthesis* that always operates in the present, negating the Goethean linking of past and present. The cut-up tends to restrict everything to the present tense. In *Rom*, Rome is not the "eternal city," but a jumble of images in which ancient and modern are juxtaposed, without history. It is well known that Goethe's expectations for Italy were shaped by the accounts of previous travelers, including his father's. He traveled to Italy with a well-formed mental map of the landscapes he would find there: he found them because they were "on the map." Brinkmann fights such preconceptions in every sentence he writes. Goethe provides no graphic maps of Italy, preferring mental maps such as the countdown of the lines of latitude analyzed above. In contrast, Brinkmann's text contains many pages with maps, most lined over with pencil to indicate routes that are never explained in the text. The effect is equivalent to the disjunctive leaps of today's internet surfing. As Adler puts it, in *Rom* "documents of history—such as the Roman forum—are for [the author] 'dead meanings,' 'rubbish'—that is, prestructural material of perception that simultaneously imposes on the perceiving subject, along with recognition of the parts, also the system within which they carry meaning" (100). As Schlösser points out, Brinkmann's use of the cut-up and his general preference for US-American literature and pop culture reflect his anxiety that the past is inescapable: Brinkmann's "fear, that his break with tradition might be in vain, that no rejection could be radical enough to escape the past, haunts this travel narrative" (122).

Rom has with justice been called an "anti-Goethe" (Grimm, Breymayer and Erhart 292). Although explicit references to Goethe are few (16, 34, 79, 115), they are pointed, as in the examples given above, and when Brinkmann turns the epigraph Goethe chose for his text on its head: "'I, too, was in Arcadia!,' wrote Goethe, as he traveled to Italy. In the meantime, this Italy has gone to the dogs and become Hell's vestibule" ("'Auch ich in Arkadien!' hat Göthe [*sic*] geschrieben, als er nach Italien fuhr. Inzwischen ist dieses Italien ganz schön runtergekommen und zu einer Art Vorhölle geworden"; 16). Brinkmann's other explicit mention shows the anxiety of influence: "I guess we should do like Goethe, the idiot: find everything to be good / unbelievable the way he continually puffed himself up in the Italian notebook: he admires every little cat-shit and has a conversation with himself about it" ("Man müßte es wie Goethe machen, der Idiot: alles und jedes gut finden / was der für eine permanente Selbsteigerung gemacht hat, ist unglaublich, sobaldman das italienische Tagebuch liest: jeden kleinen Katzenschiß bewundert der und bringt sich damit ins Gerede"; 115). Like Niebuhr and Tieck, Brinkmann's criticism does identify a crucial feature of *IR*. The later author degrades Goethean reflexivity into a simple "conversation with oneself," as if this were merely a tic on Goethe's part rather than a deliberate rhetorical strategy. Furthermore, as Immacolata Amodea points out, Brinkmann here has in fact used Goethe in order to identify his own method of composition—the whole of *Rom* is Brinkmann's "conversation with himself" (19).

In Brinkmann's text, geographic movement reveals not the "new world" that Goethe delighted in, but a sickening sameness everywhere, as Europe disappears under the onslaught of mass culture. Borders between countries are now mere phantasms: "I am traveling towards the border, as though that were a physical barrier—yes, those human pigs were so clever that they drew imaginary lines and then put all kinds of ideas into people's heads that they are now jam full of, ideas make them use words and the interests behind them to put borders around land that simply goes on and on and we think: the border, here comes something new and overpowering. Not! It just goes on" ("Ich fahre gegen die Grenze, als sei das eine körperliche Barriere—jaja, so geschickt waren die Säue der Menschen, daß sie imaginäre Striche gezogen haben und dann den Leuten eine Menge Begriffe in den Kopf gesetzt haben, die sich nun in ihnen stauen, daß sie Land, das einfach weitergeht, plötzlich mittels Wörtern und dahinter Interessen abgrenzen und man denkt: Die Grenze, jetzt kommt was Neues, das mich überwältigt, haha, geschissen. Es geht bloß weiter"; 164). What goes on and on are the processes of globalization and production of mass culture. Goethe's account submerges political boundaries beneath the telluric *differentiae* that distinguish nations from each other. Brinkmann and Goethe thus share a tendency to regard borders as phantasms and to favor telluric and geodesic differences as more meaningful. However, Brinkmann inverts Goethe's idea of the "entirely new country." Unlike Goethe, he remarks on the passing of borders, the changing of money, the diet and the sleeping accommodations: the net result of these differences is, "it just goes on." Thus, Brinkmann examines and rejects Goethe's myth of "the South" explicitly: "Do you know why I always want to go farther north? Because fewer people are there. . . . Because there is no 'South,' except as a fuzzily conceived image . . . because the south is the equator, I don't want to go there, instead to the north, the North Pole" ("Weißt Du, warum ich weiter und immer nach Norden will? Weil da weniger Menschen sind. . . . Einen 'Süden' gibts nämlich gar nicht, außer als geistig-verschwommene Vorstellung . . . denn der ist der Äquator, da möchte ich nicht hin, wohl in den Norden, Nordpol"; 187).

Like Goethe, Brinkmann shows an interest in the numerical coordinates of his position. In place of the movement through different latitudes of *IR*, however, these are stationary, and are part of a statistical analysis of the only place in his Italian journey that pleased him a bit, the Olevano Romano, about fifty kilometers from Rome, where he spent Christmas and New Year's Eve. The description—an excellent example of the "cut-up" style—shows not the distinctness, antiquity, or isolation of the place, but its complete annexation into European history and politics: "the town lies at 41 degrees, 51 minutes and 36 seconds latitude, 34 degrees, 48 minutes, longitude . . . southerly distance from Rome 1 minute 8 seconds / Rome fifty-four kilometers distant. . . . In 1944 soldiers of the German SS marched through, 1948, in apparently the first post-war election, the Christian Democrats were the top party with a mandate of 2544 votes, in second place the Communists with 137 votes, there is a picture taken on 11/21/57 showing the grinning Professor Heuss in this place next to one Olga Baldi in local dress" ("Der Ort liegt 41 Grad 51 Minuten 36 Sek. Breite, 34 Grad 48 Minuten Länge ... südliche Abweichung von Rom 1 Minute 8 Sekunden./Rome 54 Km entfernt. . . . 1944 sind hier deutsche SS-Soldaten herumgestiefelt, 1948, offensichtlich erste Wahl wieder, war die 1. Wahl wieder, war die 1. Partei die Democrazia Cristiana mit der überwältigenden Mehrheit von 2544 Stimmen, als zweite Partei die Kommunistische mit 137 Stimmen, es gibt ein Bild, das den grinsenden Professor Heuss am 21.11.57 in diesem Ort zeigt neben einer Olga Baldi in localem Kostüm"; 429).

Theodor Heuss—in party affiliation a Christian Democrat—was the first President of the Federal Republic of Germany after World War II. The allusion to the SS and to Heuss, together with the constant references to German Romantic painters and writers—such as Ludwig Tieck—who have visited the place make it seem like a southern German health spa rather than part of a foreign country. Brinkmann's cut-up is less a protest against globalization than a realization that it has been present forever, giving the lie to compartmentalized notions of national identity. Brinkmann's text represents the ultimate counternarrative to the *IR*, because the whole of its "narrative" deconstructs the narrating self, which is, as we have seen in Goethe, also the perceiving self of, above, and in landscape. Both symptom and cause of this profound alteration in subjectivity is the inability of landscape to continue functioning as a totalizing vision. To use Arjun Appadurai's taxonomy, the totalizing gaze of the landlord, of the traveler, of the colonial adventurer has dissolved into a number of nonperspectival "-scapes": "ethnoscape," "technoscape," "financescape," "mediascape," and "ideoscape." Only the first of these yields group identity (i.e., the others are heterotopic), and according to Appadurai, "the landscapes of group identity—the ethnoscapes—around the world are no longer familiar anthropological objects, insofar as groups are no longer tightly territorialized, spatially bounded, historically unselfconscious, or culturally homogeneous" (48). My analysis has shown that these themes inhabit the discourse of Goethe's *IR* from its first pages (e.g., the use of latitude lines to bound or territorialize Italy) and reinforces these differences with a continual preference for mimesis and repression of diegesis. Despite the fact that both Germany and Italy are fragmented politically, Goethe's *IR* homogenizes Italy in implicit contrast to the "formless" Germany that has yet to become a nation. Next, we will see how Gogol and Turgenev confront similar complexities in depicting "their" Russia.

Chapter Three

Diataxes of Lostness, Russian Imperial Geography, and Gogol and Turgenev

Although both Gogol and Turgenev spent more time away from Russia than Goethe from Germany, neither produced a work similar to the *Italian Journey* (the closest work to Goethe's text in Russian literature may be Nikolai Karamzin's *Letters of a Russian Traveller* of 1789-90). The prose fiction of Gogol and Turgenev is set almost entirely in the Russian empire. The heterotopic element of this Russia is not so much its contrast with other nations as the instabilities of its own territorialization and the uncertainty of its imperial ambitions. As different in their literary style as these two authors are—Gogol always humorous and surreal, Turgenev much more in the realist mode—they share a diataxis of lostness that, in my reading, symbolizes the contradictions of a Russian empire that was attempting to define itself at the same time as it was assimilating new territories and new peoples. Let us begin at the end, with the final symbol of that instability: the endless steppe.

The steppe

The sudden appearance of the steppe at the end of Turgenev's *Notes of a Hunter* (1852; "*Zapiski Okhotnika"*) receives little preparation from the geographic context of the collection's other fictions. It introduces suddenly an expansive, almost apocalyptic tone into a collection otherwise characterized by realism and a sense of rootedness in place:

> Your breathing is calm, though a strange anxiety invades your soul. You walk along the edge of the forest, keeping your eyes on the dog, but in the meantime there come to mind beloved images, beloved faces, the living and the dead, and long-since dormant impressions unexpectedly awaken; the imagination soars and dwells on the air like a bird, and everything springs into movement with such clarity and stands before the eyes. Your heart either suddenly quivers and starts beating fast, passionately racing forward, or drowns irretrievably in recollections. The whole of life unrolls easily and swiftly like a scroll; a man has possession of his whole past, all his feelings,

> all his powers, his entire soul. And nothing in his surroundings can disturb him—there is no sun, no wind, no noise. But now you've set off into the distant fields and the steppeland. You've made your way for six or seven miles along country roads and now, at last, you've reached the main road. Past endless lines of carts, past little wayside inns with samovars hissing under the lean-to out front . . . you travel for hour after hour. . . . Farther, farther! [Dalee, dalee] The steppelands are approaching [Poshli stepnye mesta]. You look down from a hill—what a view! . . . But you go on traveling, farther and farther. The hillocks grow shallower and shallower and there is hardly a tree to be seen. Finally, there it is — the limitless, enormous steppe no eye can encompass! (Freeborn 387-89; Turgenev 386-88)

As Irene Masing-Delic has noted, the journey described in this quote has as its medium less the natural features of the Russian countryside than those found in another literary text, the troika journey that ends the first part of Nikolai Gogol's earlier *Dead Souls* (1842; "*Mertvye dushi*"). As Chichikov, the anti-hero of that novel, flees the provincial society that has finally discovered his imposture, "milestones go flying by, merchants come flying at you on the boxes of their kibitkas, the forest on both sides is flying by with its dark ranks of firs and pines, with axes chopping and crows cawing, the whole road is flying off no one knows where into the vanishing distance" (*Dead Souls* 252). Turgenev rewrites Gogol by bringing greater control to the journey and by increasing the illusion that the movement is being experienced by an individual. Both journeys invoke a Russian "future fraught with endless potential. . . . Although Turgenev's epilogue to *Notes* . . . offers a more modest variant to the ecstatic coda of *Dead Souls*, there is, nevertheless, the same endless perspective and promise of endless achievement" (Masing-Delic 449). However, this feeling of potential achievement symbolized by smooth movement through the open space of the steppe has been purchased by a contrasting diataxis of constriction and lostness in movements of the narrators through the very different mental maps presented by the stories themselves.

The emergence onto the steppe represents an opening of national vision. The last words of *Dead Souls* expose Chichikov's wild ride as an allegory for that of a Russia that is constantly expanding its national boundaries: "Rus', where are you racing to? . . . looking askance, other nations and states ['narody i gosudarstva'] step aside to make way" (*Dead Souls* 253; *Mertvye dushi* 259). Both journeys smack of the picaresque genre, where, as Benedict Andersen argues, "the movement of a solitary [picaresque] hero through a sociological landscape" is powered by a "national imagination" (Anderson 30). Don Quixote became the national figure of Spain by acquiring the dust of its highways and the fleas of its taverns; in some sense, both these Russian authors are engaged in a similarly picaresque-geographical project, although neither Chichikov nor the hunter suffers from Quixotic idealism. The steppe itself symbolizes the endless possibility and "unfinished" quality of Russian history, as opposed to the more frequent concentration on limitations, enclosures, and descriptions of forest and farm. While seeming to balance the different landscapes as he does the seasons, Turgenev at the same time portrays the taxis from forest to steppe as a movement out of an enclosed, embracing space into the parlous unknown. These vicarious journeys through the Russian landscape, and the sudden appearance of the steppe, had immediate relevance to an urban, educated Russian readership of the nineteenth century familiar with the endless debate over whether Russia was a European or an Asian nation. While the debate between *slavyanofily* ("Russian nationalists") and *zapadniki* ("Westernizers") had been a constant of Russian culture for centuries, the explosion of vernacular print culture and literature in the nineteenth century had given it new urgency. In an interesting dialectic, slavyanofily imported ideas of Russian national uniqueness from the German Romantics (such as Tieck, whose pro-German critique of Goethe I presented in the last chapter). The zapadniki tended to be Hegelians, convinced that the particularism of nationhood should be sublimated in the universal Absolute and in the emancipation of humankind, which would sweep away traditional practices such as serfdom.

The journey to the open *steppe* in Turgenev and Gogol may either represent a flight from Europeanized Russia or an embracing of Russia's colonial project. Ewa Thompson claims the latter not just for these two works, but for all the canonical Russian writers of the nineteenth century, whose heroes are all "part and parcel of the colonial project" (28; see also Layton; Brower; Lazzerini). Russian geographers had figured among the early Romantic nationalists and had played a leading role in promoting Russia's "civilizing" mission to the East. Alexander von Middendorf, whose expedition to the Amur in 1842 lay behind Russian claims to that region, saw his mission as complementary to the US-American one in California and Alaska, the moves eastward and westward completing the world-historical process of European colonial expansion (see Bassin, "Russian Geographers" 119). In 1869, to give another example, Nikolai Danilevski's *Rossiia i Evropa* had appeared in installments in the St. Petersburg journal *Zaria*. With this work, Danilevskii attempted to counter the prevailing viewpoint that the Ural Mountains formed a natural boundary dividing Russia into two parts: one European; the other Asian—the way the Carpathians form an absolute border in Gogol's "Terrible Vengeance," as we shall see below (the Caucasus Mountains were yet a third geographical boundary that came to symbolize wild, untamed, Orientalist beauty in the work of Aleksandr Pushkin, Mikhail Lermontov, Lev Tolstoy, and others). The Urals, Danilevski argued, were far too easy to cross to form such a boundary:

> Russia, he suggested, represented an independent geographical world, self-contained and distinct from Europe as well as from Asia. Instead of dividing "naturally" into European and Asiatic halves, much of the contiguous land-mass covered by the Russian Empire formed instead a cohesive, unipartite "natural-geographic region." In constructing this image, Danilevski relied principally upon factors of topography and landform geomorphology. In his description, the East European plain to the west of the Urals and the west Siberian plain to the east were two adjacent sections of a single dominating landform: a vast, barely undulating lowland extending from the borderlands on the west deep into Siberia. This landmass served as a sort of unified geographical foundation of the Russian state, which was uninterrupted internally by any significant topographic features. (Bassin, "Russia" 11)

Danilevski did not prevail with his argumentation. Through the period of the Soviet Union, the Russian territory was divided into parts labeled "European" and "Asian." Gogol and Turgenev, I argue, prepared Danilevski's argument in the geographically relevant portions of their fiction. But do these impulsive, seemingly uncontrollable fictional journeys eastward imply a corresponding increase in "Asian" cultural content? Could these journeys be meant to reconcile the diversity of Russian geography and culture? Above all, do the journeys convey a single meaning, two opposing ones, or something in between? To answer these questions will require a placement of these texts in the context of other writings by each author that take account of geographical specificity, of their political and social philosophies, and of their relationship of influence.

Gogol's Dikanka and Ukrainians in blackface

Gogol-Yanovsky was born in Sorochintsy, Mirgorod district, Poltava province, in 1809. After failing in Petersburg as a poet and actor, he first achieved literary fame with two volumes of stories, the *Evenings on a Farm Near Dikanka* (*"Vechera na xutore bliz Dikanki"*), about a real Ukrainian village, Dikanka, published in 1831 and 1832. Gogol's diataxis involved the transfiguration of real geography into anagogic landscapes, phenomena into *noumena*. The themes of these stories of magic, witchcraft, incest, and murder derive from the influence of German *Kunstmärchen* by Ludwig Tieck, E.T.A. Hoffmann, and others. The choice of a rural setting and use of transliterated Ukrainian for some of the dialogue, on the other hand, hint that Gogol was practicing a "romantic regionalism" inspired by the novels of Walter Scott. Like the English and Scottish, Russian and Ukrainian national identities bear complex relationships to each other. It was Ukrainian territory that saw the origin of *Rus'* as an ethnic and political name, and the Kievan empire became the prototype for the later emergence of Muscovy. An independent Ukrainian identity emerged when its steppes and forests came under Cossack control, and particularly after the Cossacks waged a war of independence from Poland in the sixteenth century. From that time, Ukraine was split between a Polonized territory west of the Dnieper River and a Russified portion to the east of the same river. Cossacks ("kazaki") were defined not ethnically, but politically and territorially, as those who had escaped other situations (such as serfdom) to live a military life in the no-man's land of the Russian empire, where they were subject to incursions by Poles and other groups. Cossack identity was as a "border people," a category of identity theorized in the 1980s and 1990s by Gloria Anzaldúa, D. Emily Hicks, Ruth Behar, and Alfred Arteaga, largely on the basis of Chicano experience (see Michaelson and Johnson). Gogol's depictions of Cossacks in these stories and in the later *Taras Bulba* (1842) served as one of the foundations for the "Cossack myth," epitomized in Lev Tolstoi's exaggerated claim that "Cossacks have made the entire history of Russia" (qtd. in Kornblatt 13). It is well known that Gogol's view of his culture was not purely that of a "native informant"—he wrote a famous letter to his mother asking her for Ukrainian anecdotes and folklore that he could incorporate into his stories, which he wrote in Russian. These complexities impress themselves onto the diataxis of the Dikanka stories.

Judith Kornblatt has explained the necessity of the Cossack to Russia's image of itself:

> Cossacks did not "make the entire history of Russia" any more and probably even less than Daniel Boone "made" America. Despite any concrete role the Cossacks may have played in Russia's past, their mythic service has been greater. In retrospect, the Russians had no Trojan War upon which to base a positive image of themselves as victors. There were no militant Crusaders who valiantly fought a holy war. No brave pioneers battled savage Indians to make way for civilization, having successfully rebelled against their mother government. . . . The Cossacks, however, provide the Russians with an aggressive and colorful portrait of part of their past and of themselves. (13)

The frontiersman or colonizer is ambiguous, by definition operating at the edges of what has already been "nationalized." Boundaries create nations, and the Cossacks represented a kind of human boundary, the outer limit against which Russians defined themselves. Robert Maguire has argued that the theme of boundaries inhabits Russian culture from its earliest writings, such as the Primary Chronicle of the twelfth century. Russia has viewed itself continually as an orderly society threatened by disruptive forces that come from "without." Images and myths of enclosure have been created to counteract this anxiety of contagion. In Russian literature, "the myth of enclosure has proved endlessly nourishing to writers, whose explorations in turn have rooted it still deeper. None was more sensitive to it than Gogol. A 'bounded system' underlies his conception of the world in all his writings, fictional and nonfictional. In fact, he mediated it in ways that have proved decisive for many of the great writers who followed him" (Maguire 4). Boundaries, of course, are ambiguous: by definition, they belong as much to the Other as to the Self. In this mediated, liminal space, Russian intellectuals looked for the development of a national literary character. Hence the interest in Ukrainian literature in the early nineteenth century, which Gogol seized on for his first set of stories, taking advantage of his knowledge of the language and geography. As D.B. Saunders states categorically in his survey of the first critical reactions to the Dikanka stories, "the principal significance of all Ukrainian literary activity in the early nineteenth century [was] the way in which it contributed towards the development of a Slavic identity within the empire" (74). Despite the fantastic nature of these stories and the romanticized nature of their Cossack actants, the mental boundaries they draw between ethnic groups was readily recognizable to Russian readers, who saw roughly the same groups as constituting the borders of their own nation. Thus, by continually referring to Cossacks—the term occurs five or more times on some pages of text—Gogol was creating a heterotopia for his Russian readers. In an article entitled "Ukraine in Blackface," Roman Koropeckyj and Robert Romanchuk have gone so far as to compare the Ukrainian vogue in Russian literature to the popularity of blackface minstrelsy among white audiences in the US, as creations of an exotic Otherness that could be used to define the Self.

How does Gogol represent this Otherness as a mental map? As Donald Fanger points out, "Gogol uses poetic geography to unify the Dikanka tales. Ethnicity and geographical isolation function as boundaries: "The coherence of the Dikanka world can be variously approached, first and most obviously through the poetic geography already referred to. The stories all contribute to the depiction of a culturally integral community, united by language, custom, and religion; it has little contact with any other community, offering a corporate version of the solipsism Gogol later embodied in individual characters" (90). The narrators of the stories portray their Dikanka, for example, as descended from a Cossack enclave: "in those days [in the seventeenth century] almost everyone was a Cossack" (35). Crimeans, Poles, Turks, Lithuanians, and Jews are the outsiders against which the community defines itself. Notice how Poland and the Poles represent the *nec plus ultra* of experience for the grandfather of Foma Grigorievich, one of the collection's two narrators: "as he went on he came upon bigger trees than he had ever seen even on the other side of Poland. . . . On the other bank a light was twinkling; it seemed every minute on the point of going out, and then it was reflected again in the stream, trembling like a Pole in the hands of Cossacks" (83). In "Christmas Eve," a Zaporozhian Cossack, in recounting his shortcomings to the Empress Catherine, contrasts but at the same time mixes the Cossacks with other peoples: "How are your Zaporozhian troops at fault? In having brought your army across the Perekop and helped your generals slaughter the Tartars in the Crimea? . . . There are some [Cossacks] who have wives in Poland; there are some who have wives in the Ukraine; there are some who have wives even in Turkey" (128-29).

As in this quotation, most representations of this bounded space of Dikanka come replete with suggestions for its transgression or subversion. In Victor Erlich's reading, all the stories are marked by the *Leitmotif* of "abrupt, unmotivated metamorphosis" (30). For example, Gogol's romanticized geography of Dikanka blends sacred with historical time. Dikanka is a land of Cocaigne, a world turned upside down. This propensity for subverting the boundaries between real and fantastic geographies carries over in attenuated fashion into the St. Petersburg stories and also into *Dead Souls*. Nina Perlina has compared Gogol's landscapes to those of Pieter Breughel in their invocation of carnival in which people transmogrify into the cultural and natural landscapes. For John Kopper, the constant metamorphoses have a philosophical import, showing that in the Gogolian text "the noumenal world is so close to phenomenal experience that no special cognitive power like intuition is needed to make it accessible. . . . The Dikanka stories . . . carefully delimit the spheres of natural and supernatural behavior and use their intersection to generate plot" (44). We can thus speak of metamorphosis, of the transformation of natural landscape into magical events, as the governing diataxis of these stories. "A Terrible Vengeance" ("Strasnaia Mest") provides a powerful example of this metamorphosis. The story features both Cossacks and their opponents, including Poles, Tartars, and Turks. Gogol anthropomorphizes the Dnieper River, for example, the most important of the Ukraine, and one often treated as a living being in Ukrainian folktale, in a way that links it to Cossack culture: "Then its mountainous billows roar, flinging themselves against the hillside, and flashing and moaning rush back and wail and lament in the distance. So the old mother laments as she lets her Cossack son go to the war" (161). In terms of historical geography, the Cossacks' main encampment was an island in the river, and most of their raids were by boat. As the supposed protagonist Danilo floats down the Dnieper to his ancestral home, the narrator goes one step further by revealing the landscape as a living body in a prime example of the narrativization of space: "Sweet it is to look from mid-Dnieper at the lofty mountains, at the broad meadows, at the green forests! Those mountains are not mountains; they end in peaks below, as above, and both under and above them lie the high heavens. Those forests on the hills are not forests: they are the hair that covers the shaggy head of the wood demon. Down below he washes his beard in the water, and under his beard and over his head lie the high heavens. Those meadows are not meadows: they are a green girdle encircling the round sky; and above and below the moon hovers over them" (138). The symmetrical image of wood-covered hills in clear water gives rise to a myth. In this description, everything is doubled—mountains become a head, the forests hair, and so on — creating confusion as to what really lies "above" and what "below," denying the value of first-hand observation in favor of the hidden, anagogic properties of an organic and magical landscape. The structuring of the passage around a series of negations reveals, to follow Victor Erlich's reading, the Ukrainian ballads and folklore that provided Gogol his richest source of inspiration. "Yet this time the strategy of denying an element of a landscape ('forests') in order to project an eerie anthropomorphic motif ('wood demon') embodies not only the Romantic poet's urge to animate nature, but also Gogol's characteristic compulsion to displace or debunk mere sense data" (42). The passage thus anticipates the end of the story, which enlarges our perspective to reveal Danilo's persecution by his father-in-law as a vendetta being enacted on a cosmic scale. Gogolian anagogic geography leads from a cosmic, mythically holistic view of the earth back into the processes of the individual mind.

As the diegetic scale of the story grows, its mimetic force decreases, for we learn that we have been listening to a ballad sung by a blind man. As the conditions of narrative are revealed, the hitherto perplexing actions of the characters—particularly the unmitigated evil of the sorcerer father-in-law—suddenly become comprehensible. A much earlier fratricide, in which Petro had killed Ivan and his child, provokes this vengeance, in which Danilo's father-in-law (Petro's descendant) kills daughter, son-in-law, grandchild, and holy hermit, all so that he may in turn be killed and tortured by Ivan's ghost. Petro will then struggle to rise and take vengeance, but will be unable to stir from the earth, "for there is no greater torture for a man than to long for vengeance and be unable to accomplish it" (172-73). So Ivan pronounces his curse, and so it was all fulfilled: "the strange horseman [Ivan]'s ghost still sits on his steed in the Carpathians and sees the dead man gnawing the corpse in the bottomless abyss and feels how the dead Petro grows larger under the earth, gnaws his bones in dreadful agony, and sets the earth quaking fearfully" (173). As Madhu Malik points out, this last journey of the horseman to the top of the Carpathians carries out the final of a series of reversals that permeate the story: "In this story, unlike the others, the other world manifests itself not only to the boundary crosser but to all; the telescoping of space allows a view of the Carpathians from Kiev and the apparition of the mysterious horseman is universally seen. . . . We see in full measure the intrusion of the sacred into the profane world" (341). According to Miroslav Shkandrij, Petro represents Poland and Ivan the Ukraine that is attempting to throw off Polish influence only to link its destiny to that of Russia. The terrible vengeance "can be read as a symbol of Ukraine's political tragedy: it is an incomplete, conflicted, not fully functioning body politic . . . The myth describes a fundamentally divided society cursed by history, tragically torn between its Left and Right Banks, between Russian and Polish rule, and unable to form an independent entity. The suggestion is that Ukraine's ceaseless civil strife has permanently disabled it" (113-14).

Yet Gogol's mental map also assuages the agonies of these national rifts by implying that all lands covered by the horseman are Russian. According to the narrator of the story, the Carpathian mountains, at the top of which the rider sits, designate the limits of the Russian world: "Far from the Ukraine, beyond Poland and the populous town of Lemberg, run ranges of high mountains. Mountain after mountain, like chains of stone flung to the right and to the left over the land, they fetter it with layers of rock to keep out the resounding turbulent sea. These stony chains stretch into Wallachia and the Sedmigrad region and stand like a huge horseshoe between the Galician and Hungarian peoples. . . . Until you get to the Carpathian Mountains you may hear Russian speech ['russkuiu molv"], and just beyond the mountain there are still here and there echoes of our native tongue ['rodnoe slovo']; but further beyond, faith and speech ['rech"] are different" (163). Gogol's description of the Carpathians, written before any extensive geographical exploration and mapping of the region, is accurate in its essence: they indeed form a boundary (although not a barrier) between the Slavic peoples to the north, and Romanians and Magyars to the south. Here Gogol smooths over the internal national complexities of the Russian empire by enfolding them within a single circle of mountains, which magically erases the boundaries between Polish, Ukrainian, and Russian cultures that interact in the other Dikanka stories, as we have seen. The Carpathians serve Gogol as a symbol for the "imagined community" of the far-flung Russian empire. This community is founded upon the "Russian speech" supposedly found everywhere to the north of the Carpathians, and Gogol informed his mother in a letter that he was writing his Dikanka stories "in a foreign language" (i.e., not in Ukrainian).

The acculturation of landscape at the end of the story parallels its gradual petrification. As Robert Maguire has pointed out, the flowing water of the Dnieper in the early pages of the story has been reduced at its end to still, landlocked lakes, which then themselves disappear: "The last chapter, in which the blind bard explains the meaning of the events, contains not a single reference to water; he describes a *nature morte*, dominated by mountains. There is no longer a life-giving center" (12). We are but one step away from the lifeless coldness of St. Petersburg that Gogol embodied in his most famous stories: "Madman's Diary" (1835; "Zapiski Sumasshedshego"); "The Nose" (1835; "Nos"); and "The Overcoat" (1842; "Shinel'"). Concomitantly in these stories, the breadth and clarity of landscape becomes the blocked and constrained myopia of the urban. But the volumes that appeared after Dikanka, *Mirgorod* (1835) and *Arabesques* (1835), both contain transitional material. The title of the former designates a small Ukrainian town, while the latter is a collection of essays, one of which explores from a scholarly perspective the ballads that formed the basis of the Dikanka stories. Susanne Fusso tells us in her essay on the *Arabesques* that in that work "landscape is destiny," but this seems to be a departing point rather than a thesis (12). Stories such as "Two Ivans" and "The Old-World Landowners" show the decrepitude and collapse of the Cossack Ukraine, representing what in Ukrainian history is referred to as "The Ruin." According to George Grabowicz, Gogol's own flight to St. Petersburg and Russia parallels this literary resolution of the tension between Cossack and non-Cossack. The Old Ukraine is transformed into New Imperial Russia (189).

The diataxis of *Dead Souls*, written by a Gogol who had now definitively given up Ukraine in favor of Russia, is substantially different. Gone is the concern with inside versus outside, Russian versus barbarian. Instead, the journeys of Chichikov expose Russian geography as "holey" rather than "holy." In his response to the critics of his longest prose work, Gogol left a clear statement as to his intention of providing his readers with a mental map of Russia. Gogol laments, with seeming irony, the fact that Russians from all walks of life had not intervened to correct his facts, as if writing a novel were equivalent to the job of a census-taker: "The working bureaucrat could have clearly proven to me, in the sight of everyone, the improbability of the events I had depicted. . . . The businessman and the landowner could have done the same—in short, any literate person, whether he does not stir from his place or roams far and wide across all the face of the Russian land. . . . We all know Russia very badly" (Gogol, "Four Letters" 412). Here Gogol seems to be lamenting the lack of a public sphere in Russian society that would allow a broad spectrum of people to participate in a discussion of what Russia was. At the same time, his lament may be epistemological. Russia is not down on any map, mental or physical, as Melville would say. Instead, those literate enough to enter such a discussion remain separated from each other not so much by class, as by profession and specialized area of knowledge—bureaucrat, landowner, businessman. All these groups appear in *Dead Souls*, along with the serfs Gogol does not care to mention as potential sources of advice (Turgenev, on the other hand, consistently uses them as "native informants" in *Notes of a Hunter*). *Dead Souls* was seen as a critique of one group or another, even though Gogol's comments show that his interest lay in the lines of communication between groups. Like most cultures that represent borders and translations through trickster figures—the Greek Hermes, the Yoruba Legba, and so on—so Gogol represents Russian boundaries and boundary crossings through the con man, Pavel Chichikov.

The simple plot of *Dead Souls* concerns the efforts of Chichikov to buy the names of as many dead serfs as possible from the estates surrounding an unnamed provincial town, N-. N is a mediocre letter, halfway through the alphabet, as the town is said to lie halfway between Moscow and St. Petersburg. ("N.," written in Roman, is also a literary convention for unnamed towns, like "John Doe.") The list of names gives Chichikov a "virtual" estate, on the basis of which he can make loan applications to the Russian government. The opening pages show Chichikov constructing his mental map of the province: "he inquired about all the important landowners: how many peasant souls each one had, how far from town he lived" (*Dead Souls* 6). Later, the novel will take us in every direction, from the center of town to each of these estates, and then back again. With a lower-class, criminal character engaging in a random series of adventures, *Dead Souls* acquires a picaresque structure, but one that is peculiarly circular.

Let us trace the circle. One landowner, Manilov, tells Chichikov that his estate lies ten miles outside of town. Early one morning, Chichikov sets off:

> No sooner had the town dropped back than all sorts of stuff and nonsense, as is usual with us, began scrawling itself ["use poshli pisat'"] along both sides of the road: tussocks, fir trees, low skimpy stands of young pines, charred trunks of old ones, wild heather, and similar gibberish. Strung-out villages happened by ["Popadalis' vytianutye po snurku derevni"], their architecture resembling old stacks of firewood, covered with gray roofs with cutout wooden decorations under them. . . . Having driven past the tenth milestone, he recalled that, according to Manilov's words, his estate should be here, but the eleventh mile flew by and the estate was still nowhere to be seen. . . . They drove on in search of Manilovka. Having gone a mile, they came upon a turnoff to a side road, but after going another mile, a mile and a half, maybe two miles, there was still no two-storied house in sight. Here Chichikov remembered that if a friend invites you to an estate ten miles ["piatnadtsat' verst"] away, it means a sure twenty. (*Dead Souls* 18-19; *Mertvye dushi* 22-23)

We see an animation of the landscape—scenery "writes itself" (onto what?), "villages drop by"—that seems a faint reminder of the level of magic seen in the Dikanka stories, and becomes merely one of several contributing factors to the unpredictability of the journey. The confusion and "stretching," as it were, of the scale of the characters' mental maps of the region—no trick at all to turn ten miles into twenty—will become a constant of all Chichikov's visits. After leaving Manilov, he continues apparently in the same direction, with the instruction to take the third right in order to read the Sobakevich estate. However, the driver, Selifan, misses several turns: "Thinking back and recalling the road somewhat, he realized that there had been many turns, all of which he had skipped. Since a Russian man in a critical moment finds what to do without going into further reasonings, [Selifan] shouted, after turning right at the next crossroads: 'Hup, my honored friends!' and started off at a gallop, thinking little of where the road he had taken would lead him" (*Dead Souls* 38-39).

We should compare this guess on the part of the driver that a right turn should be made with the well-known dictum about labyrinths: that one should always take the left hand of a branch in order to arrive at the center. Perhaps the most famous modern invocation of this idea is by Jorge Luis Borges, who has the hero of "The Garden of Forking Paths" ("El Jardín de Senderos que se Bifurcan") take a series of left turns to arrive at the house of the man he intends to assassinate. When he arrives at the station, Yu Tsun is told by the locals that "you won't get lost if you take this road to the left and at every crossroads turn again to your left" ("The Garden" 22; "usted no se perderá si toma ese camino a la izquierda y en cada encrucijada del camino dobla a la izquierda"; "El Jardín" 106). However, left is the path of the devil in Russian and many other folk beliefs. Chichikov arrives at an estate whose owner has never ventured into town, who knows neither Malinov nor Sobakevich, and who puts up his visitors for the night. The next morning, he is led by a girl through many twists and turns on the backways, where—another animation of the landscape—"the roads went crawling in all directions like caught crayfish dumped out of a sack" (*Dead Souls* 58; "dorogi raspolzalis' vo vse storony, kak poimannye raki"; *Mertvye dushi* 62). This tangled-spaghetti image invokes what Donald Fanger calls the "flat miscellaneousness" of Gogol's mental map of Russia, one that is constantly "crystallizing into colorful incongruities" such as the specific properties of the landowners ("Dead Souls" 31).

Once onto the high road, Chichikov immediately spots an inn. It seems a reasonable assumption that the high road is one that leads straight to the town N-. Furthermore, both Manilov and Sobakevich are known to the inn personnel; so the stay with the woman estate owner seems to have represented a journey beyond the borders of the knowable community of the provincial town.

The straightness of the main roads makes perfect sense given the flatness of the terrain, confirmed at one point in the text, after Chichikov flees Nozdryov and gets entangled with a carriage containing a young girl. When she drives off, "what remained was again the road, the britzka, the troika of horses familiar to the reader, Selifan, Chichikov, the flatness and emptiness of the surrounding fields" (*Dead Souls* 91). After stays with Sobakevich and Plyushkin, Chichikov heads back for town. The repeat mention of the toll booth makes it appear that he is returning by the same road he has come: "The particolored tollgate took on some indefinite hue; the mustache of the soldier standing sentry seemed to be on his forehead, way above his eyes, and his nose was as if not there at all" (131). Earlier the tollgate had been mentioned as "striped" (18).

Gogol does not give enough details that my map of chapters 2 through 5 of *Dead Souls* can be said to be authoritative. He does give enough clues, however, to draw one (see Figure 5). The general guiding principles are worth more than the specific layout. The first is the map's reminder that Chichikov's journey consists of a series of right turns. The implication is that Chichikov, who turns always to the right rather than the left, thereby becomes lost in a labyrinth. Furthermore, there is a relationship between left- and right-handed movement and geographical orientation,

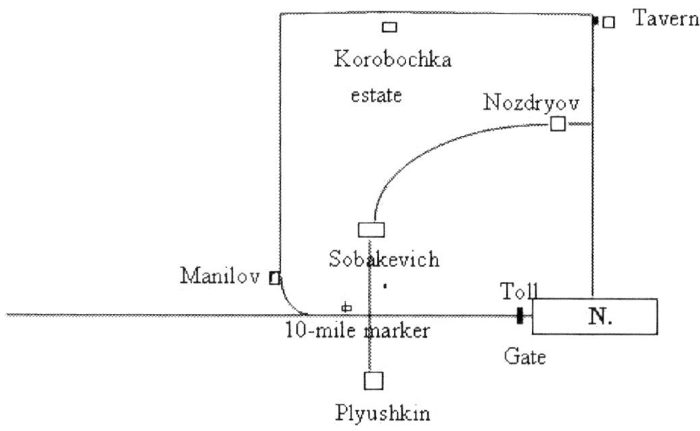

Figure 5. Map of Chichikov's serf-buying trip in *Dead Souls* 2-5

as noted by John Irwin: "The standard dictionary definition of right . . . is 'that side of one's body which is toward the east when one faces north . . . while the definition of left is 'that side of one's body which is toward the west when one faces north.' . . . Just as there is a privileged pole in each of the oppositions associated with bodily directionality . . . so there is also a privileged pole in each of the oppositions associated with geographic directionality" (161). As noted at the beginning of this chapter, the privileged pole in compass direction for Russian intellectuals like Gogol was towards the West—indeed, Gogol wrote much of *Dead Souls* in Rome. These clues make Chichikov's journey into a national allegory of Russia striving to westernize, but getting lost in the labyrinth of its own history and inertia. The failure of Chichikov's journey West leaves him the option of going East instead, in the famous *troika* ride discussed at the beginning of this chapter. A second principle is the strict boundary drawn between town and countryside, symbolized in the tollgate and in the transition from pavement to dirt. In the town, Chichikov operates by rules of society that are known in advance; in the countryside, he himself transgresses with his unusual proposition to purchase dead serfs, and the landowners also transgress, each in his or her own way, especially Nozdryov who nearly kills his guest, inverting the rules of Russian hospitality. The rules of behavior in a given situation are therefore unknown or unpredictable, the equivalent of getting lost. Finally, the map hints that diegesis in *Dead Souls* is continually drawn back into the mimesis of landscape:

> The plot here is devoured by the portraits and swamped by the details. From an engaging story the accent is shifted to the characters; but in the characters (or portraits) the accent is shifted to details of setting. The novel slips into landscape—roadside or domestic; the estates with their owners are countries, islands in an ocean; each house is sketched as a self-contained ambience with its own specific local color; as is appropriate in journeys, the geographical principle comes to the fore, character being rendered primarily as a relief arising out of the environment. . . . The formlessness and superabundance of the prose become a quality of the earth. (Terts 378-79)

Abram Terts's last sentence points to a phenomenon called "iconicity." Icons are signs that bear a resemblance to their referents—maps, for example. When the shape of the text, so to speak, comes to resemble the object it portrays, then we have iconicity. Chichikov's circular but discontinuous journey resembles the reader's own uneven and discontinuous textual traversal of *Dead Souls*. We will encounter this feature repeatedly in the works of this study (indeed, for some iconicity represented the distinction between literary and nonliterary texts).

The following passage, from the beginning of part 2 of the novel, summarizes the disorientation encountered by the reader:

> So here we are again in the backwoods, again we have come out in some corner! Yes, but what a backwoods and what a corner! . . . Everything was there together: maples, pear trees, low-growing willows, gorse, birches, firs, and mountain ash all twined with hops; here flashed the red roofs of manor buildings. . . . And this whole heap of trees and roofs, together with the church, turned upside down, was reflected in the river. . . . The view was not bad at all, but the view from above, from the upper story of the house, onto the plains and the distance, was better still. No guest or visitor could long stand indifferently on the balcony. His breath would be taken away, and he would only be able to say: "Lord how spacious it is! (258)

Gogol's prose takes us on a cinematic journey, "down the rabbit hole" and out the other side, as it were. The restrictive idea of "corner" is suddenly reversed as "everything was there together," followed by a 180-degree rotation, a second reversal where from contemplating the manor we are suddenly placed on its balcony and are looking at our own previous position. What does the pronoun "it" refer to here? There follows a description of the vista seen from the balcony, which stretches away almost to infinity, bordering on an impossible flattening of the earth's curvature which eventually hides what is distant. What is distant versus what is close: this, taken in all its different meanings, is the geographical theme of the later Gogol.

"Rus' is as much a telos of Gogol's tale as Rome is of Ovid's" (71), so say Frederick Griffiths and Stanley Rabinowitz. Gogol, like Ovid, rings the changes on forms of the epic, from mock to Christian to heroic and everywhere in between. Gogol intended *Dead Souls* to fulfill the function of national narrative, as epic had done for the Greeks and Romans, but could not match his prose to the language of the Russian epic tradition. The result is frequently called a novel, and its success prepared the way for other national narratives such as *War and Peace*, but the relation of Gogol's narrative to epic is as disorienting as the relationship of its geography to Russian reality. When Gogol's narrator invokes Rus', it is to summarize the landscape Chichikov has come to know in painful detail in Part 1: "Rus'! Rus'! I see you, from my wondrous, beautiful distance I see you: it is poor, scattered, and comfortless in you. . . . In you all is openly deserted and level; like dots, like specks, your low towns stick up inconspicuously amidst the plains; there is nothing to seduce or enchant the eye" (225). "Rus'" refers to the culture centered on Kiev that both Ukrainians and Russians see as the origin of their respective cultures. The voice is that of the mapmaker, looking down from above, from a distance. He is unable to see the many magical rabbit holes one can disappear into in this landscape. Within the labyrinth one is lost, while the bird's-eye view of it from the outside reveals nothing of its existence.

Ivan Turgenev

In an article entitled "The Hidden Dynamism of Russian Literature" ("Le Dynamisme caché de la littérature russe"), Charles Hyart has compared the *Notes* to a book published in Brussels in 1847, called *La Russie et les Russes*. The author, Nicholas Turgenev (no relation), was an economist by profession, and his work, subtitled "Tableau Moral, Politique, et Social de la Russie," has geographical dimensions. Both Turgenevs wrote from exile, and both critiqued the privileges of aristocracy and the existence of serfdom. The dynamism Hyart refers to in his title is the possibility of literature's ability to critique social institutions, and for those institutions to change as a result. Ivan Turgenev was born in 1818 on his mother's estate in Orel province. Orel lies approximately two hundred miles south of Moscow, about half-way to the Ukrainian border. Between the two authors, then, we have a nearly straight North-South axis, running from Dikanka through Orel to Moscow, and from Moscow through N- to St. Petersburg. So deeply had Westernizing tendencies taken hold in Russia in Turgenev's time that he almost had to resort to devious means to learn to read and write in his own language. Escaping from his German and Swiss tutors, young Ivan would listen to an old family serf, Fyodor Lobanov, recite Russian poetry: "the old man would recite verses by Kheraskov and Lomonosov with high feeling and flourish. Fyodor Lobanov thought it a shame that the Turgenev children were being taught by German and Swiss tutors. He alone, he claimed, was defending the beauty of Russia against this cohort of intruders. 'You are being separated from everything Russian,' he would say with a sigh, 'kowtowing to outside influences; you have turned to aliens for answers.' Young Ivan, leaning against his shoulder, was exalted by his crude, fervent speech; and it was thanks to him and his father that Turgenev learned to read and write in Russian" (Troyat 4-5). A decade or so later, Turgenev, after a period as an indifferent student in St. Petersburg, once again yielded to the inexorable pull westward, sailing for Berlin in 1838. Study in Berlin was followed by an "Italian journey." Eventually, the wealthy Turgenev spent most of his life abroad. There can be little doubt about Turgenev's *zapadnik* sympathies, given his letter in 1862 to Aleksandr Hertzen. Hertzen was a leading proponent of "Russian socialism," the notion that Russia's backwardness would allow it to bypass the negative effects of European industrialization and eventually overthrow the landowning system to engage in agrarian collectivism. Turgenev let Hertzen know that he could not share the latter's metaphor of Russia as a cousin of Europe, fresher and more attractive for her rusticity: "Russia . . . is a girl just like her older sisters—only a little broader in the beam" ("Letter" 60).

The impressions of solidity and crudity found expression in the *Notes of a Hunter* ("*Zapiski okhotnika*"), a series of short stories originally published in journals, and finally collected in 1852, the year Gogol died. They are shown, for example, in the story of the landowner Chertopkhanov, who orders a statue from Moscow for the grave of his dear friend, Ivanovich: "So it is that to this very day there stands above Tikhon Ivanovich's grave a mythological goddess with one foot graciously raised who looks with truly aristocratic disdain at the calves and sheep strolling round about her—those devoted visitors to our country graveyards" (*Notes* 326). While Gogol's invocation of the wood demon in "A Terrible Vengeance" seems romantically sincere, Turgenev focuses on the incongruity of the Greek goddess in the middle of a forsaken Russian cemetery. Western thought and ideals, including the epic tradition, are swallowed and denatured when placed in the Russian landscape. Gogol makes his description as organic as possible, so that the entire landscape comes to resemble a single living, breathing organism. In describing a cemetery, Turgenev focuses on death rather than on life, and on a peculiar death derived from the displacement of incongruous parts. The statue of Flora has been sent from Moscow, a substitute for the praying angel Chertapkhanov had ordered. This statue "had for many years decorated one of [Moscow's] overgrown suburban parks going back to Catherine the Great's time" (326). Catherine was one of Russia's greatest westernizers. It is now sent to the province in place of the desired statue. It does not look Greek, but "roccoco"; it eyes not mourners, but grazing cows, the most frequent visitors to the house of the dead. These contrasts between straightforward and ironic invocations of the supernatural, and between organic and inorganic landscape, reveal the underlying geographical bent in Gogol and Ivan Turgenev. Turgenev's knowledge of Gogol's work includes, I am arguing here, a recognition of the geographical component of the latter's fiction, to which Turgenev responds in his first texts. Both works are organized as journeys without a goal, focusing on the traveler. Whereas Chichikov's travels include mainly the "knowable communities" of the manor houses of landlords, which, as we have seen turn out in the end to be unknowable after all, the unnamed hunter supplements these with visits to peasant cottages. Symbolically, Ivan Turgenev fulfills the advice of Lobanov and escapes the Westernized drawing room for the Russian countryside. From the point of view of social class, the *Notes* traverse nearly the whole of non-urban Russian society, from the outlaws in "Clatter of Wheels" through free homesteaders like "Farmer Ovsyanikov," to landowners who "are highly esteemed and well-intentioned, and who enjoy universal respect in several counties" (182). Turgenev's laudatory obituary of Gogol, published in the *St. Petersburg Press*, brought him the unwelcome attention of the Empire's police. So, too, did the publication of the *Notes*, for which he was exiled to his Spasskoye estate. The book's heresy lay in its realistic account of the exploitation of peasants by landowners and their hirelings. Since its publication, *Notes* has been read as a critique of the landowning class and a cry of sympathy for the oppressed peasant and serf. Turgenev concerns himself with unavoidable facets of Russian existence—of all classes—in the provinces. He writes, in other words, a geography of provincial Russia. The success of Gogol had turned the interests of Russian urban readers towards their own country. As a highly educated, talented writer very familiar with conditions in the countryside, Turgenev could provide the portrait his urban readers now desired. The *Notes* are, as Leonid Grossman has pointed out, Turgenev's attempt to bring the genre of the "peasant story," as practiced by Gustave Flaubert, George Sand, Giovanni Verga, and other European authors into Russian literature. This point is made explicitly several times in the *Notes*, for example, in the following passage: "Probably few of my readers have had a chance of seeing the inside of a rural tavern; but we hunters drop in anywhere and everywhere!" (235).

The insistence of critics and readers upon the theme of oppressed peasantry in the sketches has obscured the more complex geography that Turgenev embarked upon, which includes individuals from all walks of life in relation to their environment. It has also ignored, if we are to believe Andrew Durkin and Rufus Mathewson, Jr., the elements of classical pastoral Turgenev inserted into some of the sketches—another example of the Russian landscape hiding classical motifs. Yet Leonid Grossman has shown convincingly that Turgenev's statements in this early period claim an abhorrence of political literature. Instead, he was interested in the literary realism practiced for example by his friend, Flaubert. For this reason, most of the *Notes* are sketches without well-defined plots that would reveal the mechanisms of class conflict. They are structured by space and movement through a landscape, rather than by action in the Aristotelan sense. All are told from the perspective of a solitary, anonymous individual, a landowner who hunts for pleasure. Not a single sketch takes place on the hunter's own estate, and only a few involve the hunter as an actant in the unfolding story. Most often, the hunter witnesses the action or overhears it told by others. Like Gogol's Rudy Panko, who narrates some of the Dikanka stories, Turgenev's hunter provides a unifying focus for writings that differ a great deal from each other. Some, such as "Chertopkhanov and Nedopiuskin," approach the form and length of a novella, concerned with the fate of a single individual. Others, like "Loner," are closer to short stories, with brief, surprising endings. Still others, such as "The Singers" and "Bezhin Meadows," are sketches, modeled on the French "physiologie," in which the hunter merely records a situation and the dialogue of others. It would be possible, of course, to begin analyzing the particular perspectives that Rudy Panko and the hunter impress upon the various texts that they supposedly do not produce, but merely convey to the public, but neither Gogol nor Turgenev is interested in elaborating a psychological portrait of his narrator. Instead, both narrators disappear into the surrounding context of the places they are describing. Turgenev's hunter approximates the functioning of a camera eye, with all its limitations. These are brought out most fully in the story "The Office," where the narrator lays down to rest, and then wakes up to eavesdrop on a conversation in the next room. These conversations are between the steward of an estate and the peasants who bring in their produce to sell. It becomes evident that the steward will only allow the sales to take place once a bribe has been paid. Our perception of this situation is limited to what the hunter can hear, which is broken off in many places. The peasants' point of view remains unknown to us, except inasmuch as they voice it to the dominating, corrupt official.

Walter Smyrniv is among the few critics who has attempted to correct the limitations that reading the *Notes* as social criticism has placed on their full interpretation: "These sketches represent a substantial departure from physiological accuracy. . . . Turgenev alludes to serfdom primarily in order to show the peasant's relationship to an unavoidable facet of their existence, to a feature of their reality which reveals either the realistic or idealist traits among the peasants" (75-76). Gian Piero Piretto uses a comparison with Giovanni Verga's *Vita dei campi* to reach much the same conclusion. The Sicilian peasant of the latter work finds himself in conflict with nature, whereas Turgenev's serfs submit to the laws of a nature they have helped create. Turgenev "is the first Russian fiction writer to have elevated serfs to the rank of protagonists, but . . . their presence is strongly linked to and conditioned by that of dominant nature" (311). In other words, the *Notes* show a diataxis that creates characters by placing them in landscapes, rather than by providing narratives of their past. This strategy, however, cuts both ways, and I will show how geographic description in the *Notes* carries social and ideological messages.

Gogol-like, the first sentences of the collection eschew a bird's-eye view of the Russian landscape in favor of one based on traversal: "Whoever has happened to travel from Bolkhov County into the Zhizdra region will no doubt have been struck by the sharp differences between the nature of the people ['porodoi ljudei'] in the Orel Province and those in Kaluga" (*Notes* 15; *Zapiski okhotnika* 1). While we are still at the level of "people" in general, the next sentence contrasts the peasants of the two regions, meaning that the difference between Orel and Kaluga lies chiefly in its peasantry. Orel peasants are tied to their landlords through labor, while Kaluga peasants pay quit-rent. Orel villages lie among ploughed fields without a tree in sight, while Kaluga villages are situated in woodlands. The sketch goes on to contrast two Kaluga peasants, Khor and Kalinych. Whereas Gogolian boundaries tend to differentiate Russians and Ukrainians from other peoples, however, this description dwells on the divisions within Russians. They share a language, but their "natures" are different. The texts map Russian society through a series of triangulations that mediate contrasts: landowner versus peasant (the governing contrast of the collection); free versus bound peasant; common versus "private" land; and at the end, as we have seen, forest versus steppe. Elsewhere, however, Turgenev provides direct statements of the nationalistic functions of his geography: "These were free-ranging ['razdol'nye'], expansive, well-watered, grassy meadowlands with a host of small pastures, miniature lakes, streams and large ponds overgrown at each end with willows, absolutely Russian, places dear to the heart of the Russian people ['priamo russkie, russkim liudom liubimye mesta'], like the places to which the legendary warriors of our old folk sagas used to travel to shoot white swans and grey-hued ducks" (374; 373). Here we see a very direct, simple diataxis as the narrator imagines the landscape populated by heroes of the Russian past—by implication, such a landscape is fit for heroes. Even the peasant driver is moved: "'there meadows are named after Saint Yegor,' he said, turning to me. 'And beyond 'em there'll be the Grand Duke's. You'll not see any other meadows the likes of these in the whole of Russia ['vsej Rasei']. There's real beauty for you!'" (375; 373). He goes on to describe haymaking, and fishing in the ponds. The naming after a saint, along with the adjective "razdolnye" in the previous passage, implies that these are communal meadows; "beyond" them begins the estate of a powerful aristocrat. The drawing of property boundaries inserts an element of social tension into the scene of nature, embodied in the ambiguity as to which meadow the peasant is referring to but also in his dialectal pronunciation of the word "Russia." Furthermore, the contiguity of the two territories, without communication between them—the sketch "Loner" ("Biriuk") shows the dire consequences for peasants who invade the aristocracy's private reserves to hunt or gather wood—symbolizes the social classes who live beside each other, but know each other poorly. Thus, a passage that seems to construct a Russian community through shared appreciation of landscape shows at the same time the tensions between Russian social groups whose "natures" are different. Hunting is one of the few shared activities between landowners and peasants, and most of the stories in *Notes* include a peasant guide, such as Yermolay.

By making himself a hunter, and by exposing himself to the accidents of geography and weather, the narrator penetrates the worlds of nature and of the peasant. For example, the hunter finds his way to Bezhin Meadow, in the story of that title, after he gets lost by staying out after sunset. In a seeming allusion to Chichikov's misadventures in *Dead Souls*, the hunter moves through a variegated landscape, and creates for himself a series of mental maps that attempt to provide stability: "'Ahha,' I thought, 'I'm certainly not where I should be ['sovsem ne tyda popal']: I've swerved too much to the right'" (*Notes* 100; *Zapiski okhotnika* 93). We note again the distinction between right and left as incorrect and correct ways of navigating a labyrinth, respectively. After descending the hillock and traversing damp grass, he moves along the edge of a wood, telling himself, "Now then, as soon as I reach that corner . . . that's where the road'll be so what I've done is to make a detour of about three-quarters of a mile!" (101). There is no road and the mental map is revised to account for this disappointment: "Ah, these must be the Parakhin bushes! . . . That's it! And that must be the Sindeyev wood . . . How on earth did I get as far as this? It's very odd! Now I must go to the right again" (101). Finally completely lost, the narrator walks straight ahead until he comes to Bezhin Meadow, where he finds a group of peasant boys around a fire. What is the purpose of Turgenev's spending four pages to describe the process of getting lost? While some would say that getting lost is simply an excuse for Turgenev to give yet another description of nature, it is important to recognize the fundamental difference in this description, in which bewilderment and confusion suppress any tendency towards aesthetic enjoyment. I argue that in this diegesis we can find the echoes not only of Gogol, but also of landscape painting, in which the technique of perspective symbolizes mastery over the landscape. Here, the landowner's activity of hunting has led him into a situation where he cannot achieve the perspective he desires, despite repeated attempts. He is placed, for the moment, below the level of the peasant boys around the fire, who have accepted nature—including, as we find out at the end of the sketch, death—and know where they are.

This reversal is emphasized by a curious locution when the narrator finally sits down by the fire: "I told the boys that I had lost my way and sat down among them. They asked me where I was from and fell silent for a while in awe of me ['pomolchali, postoronilis"]" (103; 98). The boys go on to tell tales of ghosts and spirits, including of their own friends who have drowned. Topographic disorientation has led to a Gogolian instability of nature and magic. As Andrew Durkin points out, this confusion can be seen as a continuation of the end of "The Singers" ("Pevtsy"): "Both stories are set in July. . . . 'The Singers' ends with the narrator, in a site of ethical confusion, making his way in the growing dark, the sound of a boy's voice echoing in the distance; 'Bezhin Meadow' begins with the narrator making his way in the darkness, confused about his physical location and finally getting his bearings when he comes upon a fire with peasant boys huddled around it. The somewhat buffoonish characters of 'The Singers' are replaced by peasant boys who are, at least temporarily, actual herders" (48). In "The Singers," the narrator's chance encounter with a country tavern, described in the most unpromising terms possible, allows him to witness an extraordinary singing match. From this unlikely scene and from the throat of Yashka the Turk, a dipper in a paper factory, "the honest, fiery soul of Russia ['Ruskaja, pravdivaja, gorjachaja dusha'] resounded" (245; 241). The music makes the narrator lose his normal orientation in time and space, placing him in a different world, so to speak. So too he loses himself at the beginning of "Bezhin Meadow" in order to find a different order of reality. As the narrator walks away from the boys in the morning, he hears the bell of the church, providing an acoustic landmark and also breaking the spell of paganism the boys' tales had put him under.

Although presented differently in terms of style, Chichikov and the hunter both lack an interiority that would allow them to remember, register, or cogitate their experiences. Chichikov's journey is undertaken with a concrete economic goal that will allow him to come up in the world, but without the proper mental map to keep him from getting lost in the labyrinth. The hunter, on the other hand, without such a goal, lowers himself to the level of the peasants. In contrast to the Gogolian hero, his getting lost and entering liminal space results in an enhanced capacity for communication, rather than a discovery of the unpredictability of human behavior. As pure sensorium, however, without memory, the hunter's experience is lost as soon as the liminal space dissolves. As with the final passage onto the steppe, a clear and enduring mental map of Russia eludes the hunter.

Return to Dikanka

An argument of this study as a whole is that literature has the power of providing mental maps to national readerships. This is borne out in the reception of Gogol. Vladimir Klimenko gives eloquent testimony to Gogol's Dikanka as heterotopia. He recounts a journey to Dikanka, Sorochintsy, and other locales of the Gogol stories. The journey begins in Moscow, where one of Gogol's readers takes the author's fictional treatment of these places as definitive: "I will begin to tell about my trip to

Gogol's native land ['rodina'] with a dialogue between a young female telegrapher and a gray-haired citizen of Dikanka (looking wise with his gray hair), which took place in the Central Telegraph Office in Moscow. 'Why are you playing with me?' The young woman asks angrily, raising her black eyebrows, 'What have you written here?' And she hands the man back his telegram. 'Senility. From so much studying here in Moscow, I must have forgotten the address,' he says embarrassed. Then he rereads everything from the very beginning. "It looks like it's correct . . . .' 'No, it's not! You are writing to Dikanka, and Dikanka is from a story. I know this well. Gogol invented it'" (Klimenko 86). We read here a repetition of the same tensions between Ukraine and Russia that motivated Gogol's writing in the first place. The Russian telegrapher considers Dikanka to be part of the Russian imaginary, a heterotopia, while the native Ukrianian knows that it was a real place before being elevated to a true imaginary one. All maps distort. In this case, Gogol's placement of Dikanka on the map of fictional places covered over its appearance on more pragmatic maps of Ukraine. His projection of a Ukrainian locale onto the Russian literary scene erased its presence as an factually existent place in the minds of Russian readers. It is like someone insisting that West Helena, Arkansas, could not exist, just because Robert Johnson sings about it. Turgenev's case was the opposite: his placement of Russian peasants "on the map," making them part of the "knowable community" of Russian literature, has obscured other dimensions of the *Notes* that address the possibility of Russia as an imagined community.

Chapter Four

Region and Revolution in Benet's *Volverás a Región* and da Cunha's *Os Sertões*

An Amerigunço in Canudos (In memoriam José Calazans)

Figure 6. Former location (1893-97) of Canudos, Bahia

The term "splendor" originally refers to a phenomenon of light, and as I look down on the watery surface edged in green that covers the ruins of Canudos, splendor in this optical sense reverberates to my eyes and blinds me. Splendor that occultates misery and violence. I am in the same geographical position that a soldier fighting for the Brazilian Republic would have taken in 1897, looking down from this height on the thousands of mud huts of insurgent Canudos, located more or less in the center of the Bahian backlands, that needed to be eliminated in order to allow the nation to survive, or so the party line went. I am standing at the borderline where, in 1897, the Brazilian Self confronted its monstrous Other. But which side was Self, which side Other? The mud-hut Jerusalem of the *sertão* occupied more or less the center of the picture (see Figure 6), and in some people's mental maps the center of Brazil—or of the earth: the center of a "white hole," a social event horizon. The sublime *sertão*, as depicted by Afonso Arinos, João Guimarães Rosa, Mario Vargas Llosa—and Euclides da Cunha. I have had the "misfortune" to visit Canudos after a season of rain, and the reservoir is full, the vegetation green. In periods of drought, as happened soon after I left, the ruins of the old church built by the Counselor and his followers could be seen emerging above the surface of the lake. Oleone tells me that I would have appreciated the apocalyptic aspect of a landscape where the desiccated plants have all turned white. Like most readers of Da Cunha's *Os Sertões* (1902; hereafter *OS*), I was moved by his agonistic mimesis, by the rhetorical power of the mental map he drew of this region. As in the Kantian sublime, I am drawn to compare my own moral and rational awareness with the terrible precariousness of life that the devouring landscape impresses upon me.

Some saw the building and filling of the reservoir I see before me now as yet another government conspiracy against the memory of Canudos. The Republic supposedly wanted to hide evidence of the massacre, perhaps the biggest in South American history. In times of drought, when the water disappears and the archaeologists invade the site, we are more likely to laugh at the failure of any attempt to cover Canudos with water. Paulo Dantas, author of an interesting fiction concerning the war, *O Capitão Jagunço*, devotes a great deal of his account of travel to the region to the building of this dam with capacity of 2,045,000 liters on the Vasa-Barris river. The actual goal was to provide a steady source of water for a region that is not technically a desert—some folk etymologies trace the term *sertão* from *desertão*, big desert—but that suffers from periodic droughts that turn it into a living hell. Indeed, the precariousness of water supply and the subsequent absence of intensive agriculture virtually define the boundaries of the *sertão*. Dantas talks with the engineer in charge of the project, who of course represents order and progress at the expense of history—the same conflicts that originally propelled the Canudos war of 1897: "This engineer, energetic and obsessed by the Cocorobó dam, is the flesh-and-blood image of progress" (Dantas 95). The word "progresso," not to mention the description as a whole, rings ironically. "Ordem e Progresso" ("Order and Progress") is the Brazilian motto, and reflects the positivist atmosphere, subservient to European philosophy and tradition, in which the Republic was founded in 1889. It was this Republic that would see fit to crush the telluric "uprising" of Canudos. No need to define "sertão"; a picture is worth everything. Or perhaps not. In penetrating the *sertão* I am at once exploring the margins of Brazil and its center: "There is only *sertão* in Brazil, which leads us to disagree with the English translation of backlands or desert. . . . It is worth emphasizing the importance of this region for Brazilian culture and literature" (Andrade Tosta 8). The *sertão* is a splendor that devours art—and more than art. Surely the Russian Empire, if I only knew it better, would show similarities to the Brazilian Empire, in its communities of men and women scattering across a wasteland—a steppe, a *sertão*—and being forgotten by the Europeanized "center," giving rise to a nation with two heads. Perhaps the racial mixing of the Cossacks could be compared to that of the Brazilian backlanders—the latter mostly white with Indian, with some admixture of Black. Yet in Goethe, Gogol, and Turgenev, this marginalized *Volk* is represented as an innate fixture of the nation's geography, while in the Americas it writes and continually rewrites each country's mental map, and many works of literature have focused on that rewriting process. For example, Canudos sprang up virtually overnight in 1893 as a place of rest and defense for the poorest of the poor, for those persecuted for their belief in a monarchy that had failed. They altered the landscape which has now been changed again, covered with water.

I spent last night in the only hotel in "New Canudos," seven or eight kilometers from the old site, to which the people were relocated when the dam was built. Canudos is not much of a tourist destination and the battle site is now officially a state historic park. The only signs of development are an archway under which our car passes on the well-rutted dirt road and a few signs to mark particular sites. One has to arrange one's own guided tours, and I have been setting up mine over the past two years with Oleone Coelho Fontes and now with Wilton de Carvalho, the owner of our air-conditioned van. Both are *Canudófilos* ("Canudos buffs"), the Brazilian equivalent of North Americans whose passion for the Civil War causes them to visit every battlefield, official or unofficial, and sometimes to participate in re-enactments of battles. *Canudófilos* retrace the steps of each of the four successive military expeditions sent against Canudos.

Two years before, I had arrived in Salvador, the capital of the state of Bahia, looking for someone to help me get to Canudos and enter the labyrinth of its history. I had interviewed José Calazans, the leading living expert on the Canudos conflict, who granted me an interview that I taped and then forgot. He also kindly invited me to a luncheon the next day, whose subject was to be a projected train trip from Salvador to Queimadas, following the route each of the four expeditions took. At the luncheon I met Oleone, a writer who, like Calazans, had made publicity of the *sertão* into his life's work. He promised to show me Canudos, and after many delays I am here finally, camera in hand. Here to do what? To see what, exactly? To extract what from the landscape, or to have it devour me in what way? From the heights I look kilometers in every direction, finding no one. Despite the lack of crowds and the reigning peace and quiet, aside from the sound of the wind, I can sense that many have been here before me: historians; novelists; even filmmakers. Like Goethe in Italy, my feelings are the products as much of the public diegeses surrounding Canudos as to the mimesis I construct from my own experience. As the embers of Canudos were still aglow after the city's destruction by government forces in 1897, Brazilian authors were already setting pen to paper in an effort to recount and explain the most significant millenarian event of Brazil, and possibly of the Americas. My impressions must be refracted through the different visions of subsequent pilgrims, from Paulo Dantas to Oleone, my current guide. But in particular I have been drawn here by the geographical imagination of Euclides da Cunha, whose *OS* became foundational not only for the history of the Canudos conflict, but also for Brazilian literature. Canudos was the event that most severely challenged the newly formed Brazilian Republic; paradoxically, da Cunha's text denouncing the massacre provided a mental map of Brazil that is still powerfully active today. In a nutshell, he attributed the war to the inability of the Europeanized elites of Brazil to understand the authentic, chthonic backlands population. The landscape of the *sertão* was as unfamiliar to Da Cunha as it is to me, and, overwhelmed by it, and under the influence of positivist theories of the influence of climate on race, he constructed the narrative of *OS* so as to explain the mentality of the backlanders within the context of the region's geography. As I look out across the flat plains to the low mountain peaks in every compass direction, and then down to the bright red hard rocky earth at my feet, as I attempt to push my gaze through the impenetrable tangle of aggressive vegetation, da Cunha's procedure seems all the more intuitive. Paulo Dantas, who unlike me felt that "in reality, the mountains were small and flat, a fact which didn't fail to disappoint me" (99), nevertheless reaffirms the accuracy and, more importantly, the evocative nature of da Cunha's "geographical fiction": "To travel through Canudos is to constantly witness the beautiful and savage aspects of the *sertão*. Euclides da Cunha achieved a frightening accuracy in his description. In his animism of the *sertão*, powered by a racial lyricism of genius, done at the expense of the obligations of science, although in accordance with his formation as a sociologist and geologist, Euclides described this flora and this 'secular martyrdom' of the earth, in flights of prose not previously attained in the Portuguese language" (127).

The reporters and historians combing the area for testimonials far outnumber the surviving *jagunços* (JAH-GOON-SOS), even of the second generation (that is, those having direct testimony of their parents), as shown in the film *Paixão e guerra no sertão de Canudos* by Antônio Olavo. *Jagunço* is a term for a rebel of Canudos, but it is applied generally to anyone who lives the traditional life of the *sertão*. After my photo-op from the high point that looks down on Canudos, Oleone, who has nicknamed me the *Amerigunço*, a condensation of *americano* with *jagunço*, introduces me to an authentic *jagunço*, João de Régis, whose father was a Canudos rebel. João lays out shell casings for us from his most recent harvest, including some from a cannon. Wilton purchases a few. João then takes us to visit the grave of Colonel Moreira César, leader of the third expedition against Canudos. The simple white cross of cement has been erected with the enthusiastic support (also financial) of Oleone, who is the author of the only complete modern biography of Moreira César. The monument is less than a kilometer from João's house, fairly far from the battlefield itself. Moreira César died from a shot to the belly as he moved towards Canudos to urge his soldiers on. In the subsequent rout of his troops by the rebels, Moreira César's body was carried several kilometers on a stretcher, but then unceremoniously dumped as the bearers felt it necessary to run for their lives. João responds with the certainty of oral tradition to our questions about the circumstances of Moreira César's final end: they threw leaves over his body and burned it. I can only imagine the exultation of the jagunços at having killed one of the toughest commanders of the army of the Brazilian republic—the anti-Christ's general.

Although he had witnessed part of the war first-hand as a reporter for a São Paulo newspaper, *Estado de São Paulo*, Euclides da Cunha (1866-1913) soon fell behind in the race to produce a full-length journalistic narrative on the basis of the Canudos war. Afonso Arinos began the fictionalization process in 1897 with *Os Jagunços*. *OS* appeared only in 1902. Winners in the race included soldiers in the conflict publishing their memoirs, and journalists such as Manoel Benício, whose "faction," *Rei dos Jagunços* appeared in 1897. Yet the popularity of *OS* was immediate, and it became the accepted mental map of one of Brazil's most marginalized regions. Da Cunha had been born in the temperate climate of the state of São Paulo. Instead of either trees or extensive cultivated fields of sugar, cotton, or coffee, the arid *sertão* offers savanna-like scrubland, called *caatinga*, that is often jungle-like in its thickness. Combine this vegetation with the abrupt declinations of the ground, and someone traversing this area inevitably feels as though lost in a maze or labyrinth. Da Cunha records exactly this feeling in his notebook of the campaign: "Today I traveled by horse a certain distance along the road to Monte Santo in the morning, and in order to satisfy a long-harbored, ardent curiosity, entered the *caatingas* for the first time. . . . I promptly got hopelessly lost in the multiplicity of flora and making my way through, the maze of the narrow veredas, I was tortured like Tantalus, a disillusioned fool ['ignorante deslumbrado']" (*Canudos* 59-60). We have seen this art of getting lost before; it bears some relation to Chichikov and to Turgenev's hunter. In Brazil, as in Russia, getting lost is not merely a cognitive dysfunction; it also reveals social and class distinctions. The unknown country is identical with the unknown, neglected people of the *sertão*. The harmless excursion foreshadows the disaster and mortality of three unsuccessful expeditions, and the immorality of the fourth

Figure 7. *Caatinga* around Canudos

that beheaded prisoners, used a prototype of napalm, and sold surviving women and children into prostitution or indentured servitude.

I experiment with getting lost, making my way past cacti and thorns into the thick underbrush (see Figure 7). Although I can hear Oleone and Wilton speaking, and Wilton's metal detector beeping as he passes it over the coat buttons of a buried soldier, I cannot see them any longer. The feeling is claustrophobic, the multiplicity of flora on all sides of me seems to entomb me. I can see across the shallow valley to the opposite hill, but not ten yards away to where my friends are excavating. I remember that the maps classify the *sertão* not as desert, but as savanna. There is plenty of vegetation, most of it eerily aggressive.

In *OS*, Da Cunha transfers the image of the maze from the flora and fauna of the region to its most noteworthy product, the Counselor's city of Belo Monte. (The official name given to the city founded on the ranch known as Canudos.) Da Cunha's description of it reminds me of German Expressionist cityscapes by Ludwig Kirchner or George Grosz: "This monstrous *urbs*, this aggregation of clay huts, was a good indication of the sinister *civitas* of the erring ones who built it. The new town arose within a few weeks, a city of ruins to begin with. It was born old. . . . There was no such thing as streets to be made out; merely a hopeless maze of extremely narrow alleyways barely separating the rows of chaotically jumbled, chance-built hovels, facing every corner of the compass and with roofs pointing in all directions, as if they had all been tossed together in one night by a horde of madmen" (*Rebellion* 144; "Essa 'urbs' monstruosa, de barro, definia bem a 'civitas' sinistra do êrro. O povoado novo surgia, dentro de algumas semanas, já feito ruinas. Nascia velho. . . . Não se distinguiam as ruas. Substituía-as dédalo desesperador de becos estretíssimos, mal separando o baralhamento caótico dos casebres fitos ao acaso . . . como se tudo aquilo fôsse construido febrilmente, numa noite, por uma multidão de loucos"; *Os Sertões* 162). Equipped with the knowledge that the *caatinga* is a maze, we are prepared to "read" man-made Belo Monte as the perfect expression of its surroundings. Euclidean diataxis is the simplest we have encountered so far, perhaps due to his ideological loyalty to positivist explanations, and is limbed in the division of *OS* into three main parts: "A Terra" ("The Land"); "O Homem" ("Man"); and "A Luta" ("The Conflict"). Each chapter is to beget the next, an explanatory mitosis. Yet, the description of the city conforms to the type of Jerusalem, outside of time, born old.

Compared with other reporters, such as Manoel Benício, who stayed for most of the campaign from June to October, da Cunha hardly cut an imposing figure. He spent much of his time behind the lines dining with the commander of the expedition, Arthur Oscar, and rarely visited the front. His dispatches to his newspaper were sporadic. Due to ill health, he left Canudos no later than 1 October, failing to witness the fall of the city four days later. If the *Estado de São Paulo* had invested travel expenses in hopes of obtaining a "scoop" in the Canudos affair, it must have been disappointed with the meager results. Five years later, however, Da Cunha had finished his monumental work, which for decades would remain the final word on what happened at Canudos, and which would transform the incident from an aberration of the birth of the Brazilian republic into an epiphenomenon of the Brazilian landscape, replacing the simple, binary answers of politics with topographic complexities. By the time da Cunha arrived at the front lines, just a few hundred yards from where I'm now getting lost, in mid-September, 1897, the government troops had succeeded in encircling Canudos, cutting off its supply lines and routes of escape for its defenders. The *jagunços* had repelled the previous three expeditions. Attacking a town like Canudos was a difficult task for a conventional army. First of all, the town lay in a remote, desert-like area of the interior of Bahia. Water and food were at a premium, and none of the four expeditions made adequate logistical plans. As a result, Canudos is one of the rare instances where the besieged inside a town were better supplied than the besiegers. Once arrived at the front, the army then faced all the difficulties of a conventional army fighting a guerilla force. The *jagunços* would fortify and dig trenches within their mud huts. Every one of the 2,000 or more houses in Canudos was a site of potential ambush. Every once in a while the legalists were able to capture a *jagunço*. After interrogation, which inevitably invoked the answer "não sei" ("I don't know") the prisoners were asked to shout "Viva a República" ("Long live the Republic"). They inevitably responded with "Viva o Bom Jesus!" ("Long live the Good Jesus"). They were then given the "red necktie"; that is, they were decapitated by having a scythe-like knife drawn across their throat in a rapid motion.

Federal troops had been besieging the movement's backland community for almost ninety days when the defenders lost their supreme leader, Antônio Conselheiro (Counselor), to dysentery on 22 September 1897. Canudos was a religious dictatorship, like Münster under the Anabaptists, and the loss of Conselheiro as the focus of resistance was a crushing blow. Shortly afterwards, a contingent of the leaders of Canudos fled the town, giving up on their cause of constructing a (relatively) just society, free from the interference of power elites. The formidable fighting force that remained did so under a saying attributed to the new commander in chief, Marciano de Sergipe: "If our Counselor has died, then I want to die too." Antônio Vicente Mendes Maciel, the Counselor, had been born in the state of Ceará in 1830. After failing at a number of things, including school teaching and marriage, Antônio disappeared from public view for three years, from 1871 to 1874. This disappearance was a form of hermitage, of solitude in the desert like Christ's or St. Anthony's. By 1874 Maciel had remade himself into the "Peregrino" or pilgrim, discarding normal clothing for a simple blue tunic, letting his hair grow, and adopting a mendicant, peripatetic existence. Maciel's only job was now to cross and recross the *sertão* of Bahia and Sergipe, stopping at the towns of the region to preach, to act as godfather for infants and marriages, and to encourage the inhabitants to personal, Catholic decency and to various works such as the building of churches or the construction of the little chapels of the *via sacra* (literally, "sacred path," alluding to the route taken by Jesus from Jerusalem to Golgotha where he was crucified) in Monte Santo.

On our way up to Canudos we had stopped for the night in Monte Santo, and I had visited the plaza where a wooden statue of the Counselor stands beside the large cannon, called the "Matadeira" ("killer"), used against Canudos (see Figure 8).

Figure 8. Statue of the Counselor and of the Matadeira Cannon in Monte Santo

It is a curious display, as intricate and contradictory as the landscape and the text I am trying to understand. Who or what is being honored? Maciel was known by several nicknames and honorifics, among them that of "conselheiro," or counselor. This title meant that Maciel was the lay equivalent of a "padre" or priest, exerting all of the priest's functions except the ministration of the sacraments. With the usual Brazilian preference for nicknames over given names, Maciel entered the annals of history under his alias of "Antônio Conselheiro." Such counselors were not unusual in the Brazilian *sertão*. They fulfilled an important function in a region where only one in three dioceses enjoyed the direction of a parish priest.

If the abolition of slavery in 1888 swelled the ranks of the Counselor's followers, the declaration of the Republic in 1889 gave the authorities new grounds for moving against him, for he could now be labeled a Monarchist and an enemy of the Republic. The truest element of this accusation was Conselheiro's decided opposition to civilian marriage, which was republican law. For him as for the majority of the backlanders, marriage could only be a church sacrament. The other major charge, that he encouraged people not to pay taxes to the republic, is probably a distortion of his real stance, which was that all taxes, from whatever source, were a robbing of the poor to pay the rich. Police forces were sent against the Conselheiristas on at least four occasions, the last and most violent in Masseté in 1893, where deaths occurred.

Conselheiro and his followers realized that they would have greater security from the incessant attacks and harassment if they could found a permanent community in a defensible position. Conselheiro named an old ranch at Canudos "Belo Monte" ("Beautiful Mountain," an allusion to the City On the Hill or the New Jerusalem). Belo Monte grew quickly in size, due not only to the Counselor's prestige and messianic appeal, but also to its favorable location near a relatively constant source of water, the river Vasa-Barris, to the relative ease with which the *jagunços* could construct their simple houses from the rachitic vegetation of the region, and above all to the availability of land and of freedom from the control by elites. The erection of a permanent settlement intensified rather than diminished conflicts with the authorities, since Belo Monte as a town acquired its own police force or civic guard, paid no taxes, accepted no republican money, and conducted no civil marriages. Prostitution and alcohol, both endemic to the ordinary life of the *sertanejo*, were forbidden in Canudos. Most irritating of all, Belo Monte was a free city, whereas every other town of the interior was ruled directly or indirectly by the large landowners, called *coronéis* ("colonels"). The only attempt at negotiating peace with Canudos came in the form of a mission of two priests, who arrived in 1895 to preach to its people. They reminded them to render unto Caesar what is Caesar's, and promised communion to all those who would agree to leave Belo Monte. Only a few took up the offer. The mission failed, and the priest's published report took on an aggressive tone that fanned the flames of hatred against the quasi-heretical, monarchist redoubt. Against Canudos, which had been founded on a remarkable cross-section of the common people of rural Brazil—peasants, freedmen, petty merchants, at least one Indian tribe, ex-bandits, and religious fanatics—formed an alliance of Brazilian elites: the Catholic Church, which was glad to see the heretical views of the Counselor stamped out; the local landowners, whose monopoly on land and labor Canudos threatened; the army, which saw itself as the defender of the Republic wherever challenged; and the polticians, who drew support from the first three groups and had nothing to gain from taking sides with a peasant rebellion.

Some days after the Counselor's death, on the first of October, a large group of prisoners, mostly women, old men, and children, surrendered to federal troops. Basing his description on others' accounts, Da Cunha nevertheless gives the impression of having been an eyewitness to the horrific spectacle of haggard, wounded women and children limping into camp in the first voluntary surrender:

> Our men viewed them with a mournful eye. They were at once surprised and deeply moved. In the course of this fleeting armistice the settlement *in extremis* was here confronting them with a legion of disarmed, crippled and mutilated, famished beings in an assault that was harder to withstand than any they had known in the trenches under enemy fire. It was painful for them to admit that all these weak and helpless ones—so many of them!—should have come out of those huts which had been bombarded for three whole months. As they contemplated those swarthy faces, those filthy and emaciated bodies whose gashes, wounds, and scars were not concealed by the tattered garments they wore—as they viewed all this, the longed-for victory suddenly lost its appeal, became repugnant to them. (*Rebellion* 471)

> Os combatentes contemplavam-nos entristecidos. Surpreendiam-se; comoviam-se. O arraial, *in extremis*, punha-lhes adiante, naquele armistício transitório, uma legião desarmada, mutilada, faminta e claudicante, num assalto mais duro que o das trincheiras em fogo. Custava-lhes admitir que toda

> aquela gente inútil e frágil saísse tão numerosa ainda dos casebres bombardeados durante três meses. Contemplando-lhes os rostos baços, os arcabouços esmirrados e sujos, cujos molambos em tiras não encobriam lanhos, escaras e escalavros—a vitória tão longamente apetecida decaía de súbito. Repugnava aquele triunfo. (*Os Sertões* 403)

Although he himself was not an eyewitness to these scenes, da Cunha uses the rhetorical device of "ocular demonstration" to make his reader one. It is a fiction; we readers are moved by the passage due to the creation by the narrator of sympathy, mourning, and compassion in the hearts of soldiers who may have felt none of these things—certainly they showed the inhabitants no mercy.

On 5 October the last defenders of the town were eliminated. The army gathered in front of the "new church" that the *Conselheiro*'s followers had built and the cannonballs had leveled. Miraculously, a large cross in front of the "old church" had not been hit and is still preserved today in a small chapel. The band played the Brazilian national anthem. Then the kerosene was applied everywhere, to corpses and houses, and the town burned to the ground. And so the directive of the forces of civilization was fulfilled: stone should not remain on stone. A town which some claimed to have been the second-largest in the state of Bahia was returned to wasteland. In the period of the war itself, Canudos was interpreted contradictorily as a redoubt of religious fanatics and bandits and as a counter-revolutionary state-within-a-state financed by the substantial number of monarchists remaining in Brazil—when Canudos was founded in 1893 the Republic was only four years old. Almost immediately after the war, a counter-memory of Canudos began to be constructed by authors such as Afonso Arinos, Manoel Benício, and Euclides da Cunha. Canudos is not just something that happened, but a literary *topos* that has been treated from a variety of ideological and aesthetic perspectives. The war of bullets produced a war of words, in which basic concepts of *brasileiridade* ("Brazilianness"), modernity, and social justice have been debated under the pretext of getting the Canudos story "right." The messianism of Conselheiro has been transferred to the single-minded mission of finding the ultimate truth of Canudos. Lori Madden expresses accurately the trajectory and interest of Canudos historiography when she writes that the Canudos conflict "has stimulated the imagination of diverse writers of alternative points of view since it affords evidence to be viewed as a political rebellion, a civil war, a problem of ethnicity, a messianic movement, a social movement, and other phenomena. It has become a mirror to the manipulations of its interpreters to such a degree that Canudos historiography, studied over time, tells a story of the evolution of ideas" (6). Some of those ideas involve territory and geography. *OS* certainly is one of these, constructing an image of Canudos as a monstrous city, born old, the true imaginary place as nightmare.

Like Goethe's *Italian Journey*, *OS* is a polyglossic work bordering a number of different genres. As Franklin de Oliveira points out, an entire critical tradition, from José Veríssimo to Afrânio Coutinho, has defined *OS* as a work of fiction, with generic categories varying from novel to epic (15). In addition, M. Cavalcanti Proença has provided an extended analysis of the work's conformity to tragic conventions (251-67). The corpus of Greek tragedy rather than the Bible shapes much of the prose of *Os Sertões*. Proença justifies his undertaking by citing a passage in which Da Cunha self-reflexively renders a scene of the Canudos war as though it were a staged play: "all the huts adjacent to the engineering commission constituted an enormous theater pit from which to view the drama that was taking place. Focusing their binoculars through all the crevices in the walls, the audience stamped, applauded, shouted bravos, and hissed. In their eyes the scene before them—real, concrete, inescapable—was a stupendous bit of fiction which was being acted out on that rude stage to the sinister glow of leaping flames" (*Rebellion* 432; "todas as casinhas adjacentes à comissão de engenharia formavam a platéia enorme para a contemplação do drama. Assestavam-se binóculos em todos os rasgões das paredes. Aplaudia-se. Estrugiam bravos. A cena—real, concreta, iniludível—aparecia-lhes aos olhos comos e fora uma ficção estupenda, naquele palco revolto, no resplendor sinistro de uma gambarra de incêndios"; *Os Sertões* 368-69). The passage demonstrates a major difference between *OS* and its factional predecessors: its degree of self-consciousness. Da Cunha not only fictionalizes the events of Canudos, but reflexively points to his own fictionalizations or, in the passage above, justifies them as inseparable from the attitudes of the historic actors themselves. This self-consciousness, as Luiz Costa Lima has pointed out, stems from "the incompatibility between literary models for *Os Sertões*, such as tragedy, and the author's ideology of positivism that negates the content of such genres" (185). Da Cunha spends so much of his prose describing the landscape of the *sertão* in order that it may carry the weight of explanation, thus eliminating notions of fate and destiny that would enable him to write tragic fiction.

Authenticity has been a focal point of critical debates concerning the relative value of Da Cunha's text, foregrounding the text's participation in the cycle of public interpretations and private perceptions. *OS* emerges from an extensive and often contradictory intertext. The author was only present for one month of an extensive campaign, and did not witness the fall of Canudos. He relied on friends for various data: Arrojado Lisboa gave notes on Brazilian geology; Teodoro Sampaio helped with geography and history; the medical doctor Nina Rodrigues provided psychiatric evaluations of the Counselor and his adherents. The book is, in a sense, a collaborative effort by a number of Bahian intellectuals. "*Os Sertões* would become, in fact, a mirror of paradoxes, reflecting accurately its constantly shifting background, sometimes looked at thus, by those who see themselves in this mirror without eyes for seeing the whole picture: for geographers, a great creative writer ['escritor']; for some writers a notable historian; too much of a geographer for the historians, too much an historian for the geographers" (Souza Andrade 65). *OS* distinguishes itself precisely through its foregrounding of diataxis, of the immense rhetorical labor of the author, as evidenced in his muscular (Portuguese: *hirsuto*, lit. "hairy") style. Euclides, as Gilberto Freyre points out, was attracted by "the angularity and boniness, the muscularity of the ascetic or dryly masculine aspects ['relevos'] of the sertão. He emphasizes the most tough and angular of the aspects in words that are themselves tough, almost without fluidity and, as it were, asexual" (xv-xvi). *OS*, then, like Gogol's *Dead Souls*, is an iconic text, that is, it resembles the landscape it describes. The laborious syntax, the labor of reason working on the irrationality of landscape, makes the traversal of *OS* resemble the arduous traversal of the *sertão* itself. As Maria Tai Wolff put it, "the text . . . begins to resemble the *sertão* itself. In the end, the 'tormented' landscapes of the latter take over—the people who live or enter there, the war that is conducted there, and also the text which tries to describe it. To the *sertões* of the title must be added this one: the text-*sertão*" (61).

Distinctive in da Cunha's style, for example, is the alternation of long sentences with short ones, a stylistic device which frames the alternation between analysis and synthesis. The result is a "hydrograph" of the *sertão*, comparable "to one of its rivers, whose waters do not form counter-currents or backwaters where they would shape the land in a mild and serene flow. Rather, they go leaping over obstacles, jump madly into the abysses; advance or retreat in sudden stops, trying to contain themselves in the curves of their beds" (Corrêa 7). The Vasa-Barris that I see and photograph near Canudos and near Uauá, the scene of the first battle in the Canudos war, shows all the signs of this violence. There is no defined flow when I see it. The water pools in small tanks made by the boulders that are washed downstream when a storm breaks. Da Cunha calls the Vasa-Barris a "rio sem afluentes" (*OS* 19; "possessing no sources"; *Rebellion* 18), meaning that it will begin flowing at whatever point rainfall occurs. Now I see exactly what he means. There are no steady streams, only the alternative of raging torrents or, as now, a series of stagnant pools interrupted by stretches of sand and granite (see Figure 9). An unpredictable landscape, a landscape unto itself, one that I cannot "read" out of my experience with California, Germany,

Figure 9. The "River" Vasa-Barris

New Hampshire, or Pennsylvania. Similarly, the use of the gerund creates a profile of a landscape where great blocks of granite arise from vast, flat expanses: "The artist of the word rescues the geographer to give us an acoustic image of the accidents of the terrain. The gerund creates a quick cut in the author's thoughts, who briefly abandons the descriptive sequence, making a pause for a dynamic-visual representation of the geographic features" (Corrêa 11).

Beyond its iconic function, the agonistic language of *OS* also reflects the double-consciousness of da Cunha. Beginning his text from the position of a confirmed supporter of the Brazilian Republic, his research and experiences led him to discover the deeper reality of Brazil that republicanism imported from Europe disguises but does not erase: "being ethnologically undefined, without uniform national traditions, living parasitically on the brink of the Atlantic in accordance with those principles of civilization which have been elaborated in Europe, and fitted out by German industry, we played in this action the singular role of unconscious mercenaries" (*Rebellion* xxx; "etnologicament indefinidos, sem tradições nacionais uniformes, vivendo parasitariamente à beira do Atlântico dos princípios civilizadores elaborados na Europa, e armados pela indústria alemã—tivemos na ação um papel singular de mercenários inconscientes"; *Os Sertões* xxix). Having climbed down from the heights of Alto da Favela into the air-conditioned comfort of Wilton's Korean-made van, Oleone reveals to me the mystery of Canudos. Understanding Canudos means confronting three crazies ("três loucos"): the first is the Counselor; the second is Moreira César; and the third is Euclides da Cunha. I look around the van, then look him in the eye: that makes six of us. And perhaps there is a seventh: Juan Benet.

Region and revolution in da Cunha's *Os Sertões* and Benet's *Volverás a Región*

In 1967, the engineer Juan Benet published a novel entitled *Volverás a Región* (hereinafter *VAR*; trans. Gregory Rabassa as *Return to Region*) that created a furor at the time and has puzzled readers ever since. *VAR* represented a shift in the postwar Spanish novel away from realism towards a more discursive, self-conscious kind of fiction, and a startlingly original approach to the still ubiquitous theme of the Spanish Civil War (1936-39). It also represented a remarkable adaptation of the themes and styles of American authors da Cunha and Faulkner to the locale of Spain. The double influence crossed time boundaries; Faulkner's best work is from the 1920s and 1930s, while da Cunha's *Os Sertões* was published in 1902. To give but one example, all three authors link the marginality of their regions to failed rebellion. A polyphonic reading of *VAR* against both da Cunha and Faulkner would have great merit; however, in this study Faulkner will be compared with another American author, José Lins do Rego. In this chapter I examine the revolutions in da Cunha's and Benet's narratives, as well as the devolution of regional mental maps from Brazil to Spain.

Región's two main meanings sandwich the idea of nation: region can be either a building block of nation, or a supranational geographical region, such as the Americas. In either case, there are tensions between regional and national identities and ambivalences as to whether an area is a region or a nation. Such tensions and ambivalences in the mimesis drive the diegeses of the texts examined in this chapter. The predilection for identifying regions as something set apart from the nation as a whole depends first of all on an acceptance of the ability of nation to provide a unifying set of norms, beliefs, and practices. Regions are then defined along their deviations from those norms in a process of mental mapping: "what we choose to think of as regions are mental constructs bearing little relation to geography or history; they function as preserves where we discover savages or approximations of the natural man, according to our needs" (Miller 3). Jim Wayne Miller's description of the cultural function of region is applicable to both *VAR* and *OS*. Both these works chronicle the central tragic myths of their respective nations—the Spanish Civil War (1936-39); and the 1896-97 campaign against Canudos, respectively—as a conflict between center and region.

For purposes of direct comparison I limit myself to *VAR*, the best-known of Benet's "Región cycle" which, similar to Faulkner's Yoknapatawpha novels and stories, revolves around a single, mythical place and uses the same characters repeatedly. Following the publication of *VAR*, Benet became for a time the most important prose writer of Spain, but with as many detractors as supporters. He was accused of being an obscurantist, an elitist, and an imitator of Faulkner. Yet much of the difficulty of Benet's fiction arises from the fact that diegesis in his works is not simply a chain of events, but a constant moving back and forth between the narration of historical events, the topographic situation of those events, and an exploration of the impact of events and geographies on the mental states of the characters. As a water-projects engineer with an intense interest in hydrology and mapping, Benet constructs plot recursively, the way he would describe a river: each tributary must be followed to its source, each tributary of the tributaries followed to its source, and so on. This diegetic recursion follows the contours of the mountains and rivers of Región. The name of the main river, the Torce, seems to derive from the Spanish verb "torcer," meaning "to twist," just as the plot of the novel twists and turns. In da Cunha's work, on the other hand, the twists and turns are less in the diegesis than in the language and in the meta-narrative of a Brazilian nation in conflict with itself. In both works, the torture (derived from the same Latin root as "torcer") is also located in the mimesis of humans in eternal conflict with the earth. Thus, the plot of *VAR* is not easily retold. The main action is the return of Marré Gamallo to the town of Región, where she interviews Doctor Sebastián in order to extract information about her own past. Her father, Colonel Gamallo, had led the Francoist forces against the Republican defenses of the town. His motives were more personal than ideological: years before, he had played a card game at a local casino where he had lost both his money and his lover. At the end of the interview, Marré tries to leave town in her vehicle, but is apparently assassinated by the guardian Numa who prevents all from leaving Región, while the doctor is killed by his only patient, a demented war orphan who believed that Marré was his own mother come to reclaim him.

In another work of the Región cycle, Benet's narrator has specified the differences between war at the local and national levels: "because it was local the war in Región was conducted quite differently from how its authorities and directors thought it would be. With only a few exceptions—for example, the peasant revolts—war in a narrow and backward district always comes from the outside, is more of a gift from the government and the capital, an irruption of modernity into the kingdom of anachronism. Without anything having occurred inside its borders, suddenly one July morning the district found itself at war" (my translation; "por ser local la guerra en Región era muy distinta a como la pensaban y consideraban los responsables y dirigentes. Salvo muy pocas excepciones—salvo las revueltas campesinas, en último término—la guerra en una comarca apretada y atrasada viene siempre de fuera, es un regalo más del gobierno y la capital, una irrupción do lo modero en el reino de la anacronía; sin que nada nuevo haya ocorrido dentro de sus límites de repente la comarca, una mañana de julio, se encuentra en guerra"; *Herrumbrosas lanzas* 1, 74). In other words, the supposed borders of the region with which its inhabitants had hoped to hold off modernity, indeed even its separation as a region in relation to a distant "center," turn out to be illusory. Similarly, the labyrinthine, discontinuous, and recursive narrative discourse of these two works implicitly questions their ability to construct their regions or to separate them from the "center" to which they must succumb.

True imaginary regions

Benet has given several different accounts of the origin of Región, his true imaginary place in Spain analogous to da Cunha's *sertão* and Faulkner's Yoknapatawpha county. Benet's Región is at once a town (oxymoronically), a region (eponymously) and Spain (allegorically). The fiction of Región arose simultaneously with hydrological fantasies as responses to the "reality principle" of Benet's training as an engineer. Benet was born in Madrid in 1927 and remained a *madrileño* all his life, at least in spirit. During the Civil War, his father was killed, and his family had to flee to San Sebastián in Basque country. Then, upon receiving a doctoral degree from the Escuela Especial de Ingenieros de Caminos, Canales y Puertos (Technical School for Engineers of Roads, Canals, and Ports) in 1954, Benet was assigned to projects in Northwestern Spain, where he lived for ten years—a Homeric span. Benet has recounted his stay in Pamferrer and other towns of León, and his early writings, in the short essay, "Región." It was there, in exile, as it were, that Benet began the "Región cycle," with the collection *Nunca llegarás a nada* (1961) and the novel *Volverás a Región*. Hence, the similarities with da Cunha begin with the fact that Benet is someone living at the center looking at another region of his country as a periphery (this perspective may be considered as the reverse of the work of Lins do Rego and Faulkner). From 1958 to 1989 he was the chief director of public works in Spain, writing fiction in his spare time. Further works in the Región series include *La otra casa de Mazón* (1973), *Una meditación* (1985), *El aire de un crimen* (1980), "Numa, una leyenda" (1981), and *Herrumbrosas lanzas* (1983-86). In addition to fiction, he also published non-fictional (?) works, such as the *Vistas de las obras del Canal de Isabel II fotografiadas por Clifford* (1988) and *El camino del Guadiana* (1991). Benet died in 1993.

The Región series takes place in a single, geographically circumscribed region of northwestern Spain. Benet has given several different accounts of the origin of Región, the fictional area of Spain of which he is sole owner and proprietor. One, alluded to above, is that he overlay the mental maps of da Cunha onto Spanish reality. Another is that the fiction of Región arises simultaneously with hydraulic fantasies as responses to the "reality principle" of his training as an engineer. Construction prohibits innovation, and the reality of success and failure deflate theory. At the same time, only imagination can move history forward. Hence, Benet poured his imagination into a novel, *Volverás à Región*, during whatever free time he had on the construction projects: "Thus, in order to operate completely according to my own desires, in order not to be obligated to report to anyone, in order not to have to pay the extravagant price exacted by the strict reproduction of reality, in order to be master and sovereign of a territory ['territorio'] belonging to me exclusively, I invented for my book an imaginary place ['paraje'] that I called Región. Nothing would have delighted me more than to be a kind of hydraulic tyrant ['tirano hidráulico'] of that place ['paraje']" ("El Agua en Región" 72). Benet continues this talk, delivered to the Center for Hydrographic Studies in Madrid, with some examples of his hydraulic fantasies, such as the "hairy canal," in which fur lining the bottom and sides would help speed the water along.

Benet's fantasies are not completely divorced from reality. He has recounted his idea of connecting a number of reservoirs on the south side of the Cantabrian range through underground tunnels. He actually did some surveying for this project: "I even managed to visit the terrain and record some very summary data with a view toward a project for linking the Ebro river with the Órbigo by connecting all the reservoirs of the south side of the Cantabrian mountain range, almost all parallel to each other and with similar characteristics. Ebro, Cervera, Camporredondo, Riaño, Porma y Barrios de Luna, con el hiato de Bernesga y Torio . . . would together create a connected water reserve of incalculable pòtential" ("El Agua en Región" 72). Benet's "imagineering" includes a complete listing of the places that would be linked in this project, each name rolling off his tongue with a poet's relish. *VAR* abounds with similar lists of fictional names, such as Región's main peaks, "the angular Torres, el Monje, Malterra and the bald mountain of San Pedro" (*Return* 32; "El esquinado Torres, el Monje, El Malterra y la pelada montaña de San Pedro"; *VAR* 40). For Benet, the most fascinating aspect of hydrology is the finding or construction of subterranean linkages that provide an image of the porous boundary between regions. Water, like war, can erupt suddenly in a region that had felt itself isolated from its effects. Indeed, Ronald Rapin has explained Benet's use of "Región" as an attempt to illuminate the shifting boundaries and loyalties of the Spanish Civil War: "residents of Región live in a province whose borders fluctuate continuously and geographically as the region 'implodes' inwardly on itself due to the civil war" (Rapin 542).

We can read Benet's description of the fictional rivers Torce and Formigoso, on which Región is located, as yet another incarnation of his hydrographic imagination: "Out of the strangulation comes the birth—and the divorce—of its two main rivers: toward the east the Torce, a leaping brook that with stumbling begins a brief and errant trajectory that only at its confluence with the Tarrentino brook (an impressive, somber, and blackish formation of quartzite in vertical planes and sawtooth edges), takes care to correct itself so as to render its waters where they are needed. And toward the west the Formigoso, which, in comparison to its twin, from its birth observes a straight, disciplined, and exemplary behavior so that, with no need for teachers, it will come of age according to the model established by its parents and new inductees" (*Return* 31-32; "hacia levante el Torce, un arroyo altarín que inicia con malos pasos una breve y equivocada trayectoria que sólo a la altura de la confluencia con el arroyo Tarrentino (una impresionante sombría y negruzca formación de cuarcitas en planos verticales y bordes en dientes de sierra) se cuidará de emendar para que rinda sus aguas donde son necesarias. Y hacia ponente el Formigoso que, en comparación con su gemelo, observa desde su nacimiento una recta, disciplinada y ejemplar conducta para para, sin necesidad de maestros, hacerse mayor de edad según el modelo establecido por sus padres y recipendarios"; *VAR* 39-40). Like gems within a clay matrix, fragments of scientific, rational discourse are embedded in a mythical story that anthropomorphizes these "twin" rivers. The discourses flow side by side but do not really meet each other. Likewise, the professionally drawn maps that "illustrate" the Región series provide an illusion of precision that is gainsaid by a text that consistently contradicts itself. Benet's diataxis, then, like da Cunha's, reveals engineering principles while at the same time deconstructing these.

The confluence of *Volverás a Región* with other fiction

Malcolm Alan Compitello has pointed out that although many critics have cited Faulkner as an influence on Benet, the structure and density of geographic description in *VAR* make it more closely resemble *OS*. Compitello has analyzed the parallels with da Cunha and Randolph D. Pope the parallels with Faulkner. Furthermore, Benet himself has written far more on his discovery of da Cunha than on his reading of Faulkner. He devoted an entire essay to describing how he came across da Cunha's book while attempting to communicate with his Galician workers. Galicia is a region of Spain bordering northeastern Portugal, with a dialect resembling Portuguese. Unable to find a textbook in Galician, he was given *OS* to read as his introduction to Portuguese. According to his own account in "De Canudos a Macondo," he shut himself into his construction-site shack and began devouring this work written in a foreign language he did not yet know how to read. In describing this bizarre scenario, Benet seems to re-enact playfully da Cunha's writing of *OS*, carried out in a little shack during the engineer's free time on a construction project in Brazil. In Benet's actions we already see the conflation of the text under construction with the region to be described. Book, shack, or region, each functions as "framed or hermetic space, endowed with its own laws and specific symbols" (Rosa 588).

Readers familiar with the language of *OS* will find echoes in *VAR* of the idea of tortured landscape and even of some of its vocabulary: "at the level of the inn of El Quintán, the vegetation becomes sparse and thin, low clumps of oak and mounds of whitish soil in shapes tortured by the strong March gales, up to the place where for more than eight miles there is no other shade except that of an old masonry bridge under which—except on torrential days when a tumultuous, deafening, and red freshet passes—there runs a thread of water that almost all year long can be held back with your hand" (*Return* 2; "a la altura de la venta de El Quintán la vegetación se hace rala y raquítica, montes bajos de roble y albares de formas atormentadas por los fuertes ventones de marzo, hasta el punto que en más de cinco kilómetros no existe otro lugar de sombra que un viejo pontón de sillería por donde—excepto los días torrenciales que pasa una tumultuosa, ensordecedora y roja riada—corre un hilo de agua que casi todo el año se puede detener con la mano"; *VAR* 8). This passage from the novel's beginning combines several descriptions in *OS*, but most particularly the hydrographic description of "great boxlike river beds which are dry most of the time, being filled with water only during the brief rainy seasons. Obstructed for the most part by thick layers of stone blocks, between which, save in the case of sudden swellings, thin streams of water may be seen trickling" (*Rebellion* 14; "abertos em caixão, os leitos as mais das vezes secos de ribeirões que só se enchem nas breves estações das chuvas. Obstruídos na maioria, de espessos lastros de blocos entre os quais, fora das enchentes súbitas, defluem tênues fios de água" *Os Sertões* 14). Benet's mimesis rests not only in his experience of northern Spain, but equally in his experience of da Cunha's textuality. More important than the actual borrowing of descriptive features, however, is Benet's adoption of da Cunha's dramatism in depicting an active landscape that struggles with and defeats its own inhabitants, not to mention invaders from the outside.

In his article on "Región's Brazilian Backlands," Compitello gives four points of similarity between *VAR* and *OS*: 1) similar geographical features, such as extreme aridity; 2) a similar ideology of protest—da Cunha's against a republic which had in the end shown itself to be a deeper and more destructive barbarism than that of the backlanders, Benet's against Francoism; 3) the central characters of Antônio Conselheiro and Colonel Gamallo as examples of misguided fanaticism with roots in disturbed personalities; and 4) the order of presentation in the two works, from geography/geology to climate to the effects of the first two on humanity. For example, the similarities in the structure of presentation are most striking, given da Cunha's clear motivation for presenting his material in that order, and the lack of such motivation in Benet. Given that da Cunha began *OS* with a determinist-positivistic, Darwinist model of scientific truth in mind, the order in which he presents his material follows the law of cause and effect: topography and climate have created the *sertão* and cut it off from the center of Brazilian geography, a neo-European center located on the coastline (as if in recognition of the paradoxes of having a littoral "center" of the country, the Brazilian government moved its functions to an artificially constructed capital, Brasília, during the 1960s). The geographic isolation of the *sertão* has in turn given rise to a way of life devoid of order and progress, which has allowed the backlanders to accept the insane, antirepublican rantings of Antônio Conselheiro as gospel, which, combined with the lack of self-confidence in the new Republic, has propelled the conflict and the ultimate conflagration in Canudos. In *VAR*, on the other hand, Benet symbolizes the conquering of Región by the rebel (Francoist) troops as the revenge of Colonel Gamallo against an area which had earlier deprived him of his lover and his money. Even Gamallo's personal past accounts only for the *élan* with which he pursues his campaign, not for the eruption of the conflict, nor for the tenacity with which the Republicans hold the city, nor for the mysterious actions of the other main characters. There is no cause and effect in *VAR*, only fate and myth equally guilty of the free distribution of such causes: "The truth only rarely shines forth" ("Rara vez la verdad alumbra"), as Benet has said elsewhere, referring simultaneously to his fiction and to the Spanish civil war ("Breve historia" 160). Benet's motivation reads as the mirror image of da Cunha's, for he wrote *VAR* several decades after the end of the Civil War, and in response to the host of more realistic, quasi-documentary novels which had appeared in the intervening thirty years, and which had "produced few changes" in Spanish public opinion (Thomas 209).

Benet's reticence could be interpreted as a parody of one of the most famous and respected Spanish postwar novels, Rafael Sánchez Ferlosio's *El Jarama* (1955; trans. Margaret J. Costa as *The River*). *El Jarama* invokes the civil war above all in its title, that river being the site of one of the war's most famous battles. Ferlosio begins his novel with a long geographic description, which he places in quotes in order to confirm its facticity by showing that it has been drawn from an encyclopedia. The description of the Jarama brings us from the river's source to the geographic location where the novel's action will take place: "I will describe these rivers briefly and in order, starting with the Jarama, whose source is to be found in the gneiss of the southern slopes of Somosierra, between the peaks of Cebollera and Excomunión. . . . From Talamanca to Paracuellos the river flows through various gullies as far as Puente Viveros, where it crosses the Aragón-Cataluña road, sixteen kilometers outside Madrid" (*The River* 17; "Describiré brevemente y por su orden estos ríos, empezando por Jarama: sus primeras fuentes se encuentran en el gneis de la vertiente Sur de Somosierra, entre el Cerro de la Cebollera y el de Excomunión. . . . De Talamanca a Paracuellos se pasa el río por diferentes barcas, hasta el Puente Viveros, por donde cruza la carretera de Aragón-Cataluña, en el kilómetro diez y seis desde Madrid" *El Jarama* 7). This description, which Ferlosio encloses in quotes in order to render it authoritative, inserts the events of the novel, all of which take place within a single day and are described in a hyperrealistic style, within a continuous geography and a continuous time symbolized by the flowing of the river. The perspective adopted in this paragraph is literally inhuman, available to no one, synthesized from a thousand single glimpses of the river (in contrast, the text of the novel itself refuses to transcend the characters' individual perspectives). The last paragraph of the novel continues the quotation by describing the further course of the Jarama as it flows away from Madrid. The events of a single day, whether tragic or trivial, are recuperated through their insertion into a recognizable geographic order. They become a parenthesis or hiatus in the continuity of time.

One can imagine Benet reading this introductory paragraph and setting out to parody it in the description of the twin rivers cited above, or in these opening lines of *VAR* : "It's true, the traveler leaving Region who wishes to reach its mountain range by following the old king's highway—because the modern one has ceased to be such—will finds himself obliged to cross a small, high desert that seems endless" (*Return* 1; "Es cierto, el viajero que saliendo de Región pretende llegar a su sierra siguiendo el antiguo camino real—porque el moderno dejó de serlo—se ve obligado a atravesar un pequeño y elevado desierto que parece interminable" *VAR* 7). Here we do not have the satellite view, the bird's-eye, cartographic perspective of Ferlosio, but rather a description which fails to hold itself outside the experience it is describing. Rather than using language to describe, as Ferlosio does, Benet turns language upon itself to reveal its inconsistencies with reality and experience. The departing traveler *arrives*; the *small* desert seems to never end. The modern road is no longer the king's road, because there is no more king, only a *caudillo* and a country caught in a death struggle with itself. The very first words of the novel, "es cierto," introduce a text in which nothing is certain.

Mapping Región

In José Rivero's formulation we read that "Región and its cartographic representation by Juan Benet superimpose two plots: physical; and political . . . in order to synthesize the fusion of these two time-frames, geological and social, physical and political, in a single whirlwind of time, narrated and mapped. . . . Región is mapped and narrated, drawn and related, thought and constructed on the basis of a single creative impulse of time and space" (153). Thus time and space belong to diegesis and mimesis, respectively, and it would seem that Benet's diataxis reveals the disjunctions between these two plots. A comparison of Benet's texts with any of the several graphic maps he produced for Región, such as the one as one page 206 of the first volume of *Herrumbrosas lanzas*, reveals numerous inconsistencies, which means either that following the map would soon get the traveler hopelessly lost, or that following the text would—or what is more likely, that both instruments ostensibly intended for capturing reality in fact create more room for equivocation, doubt, and (mis)interpretation (on this, see Margenot). Various readers of *VAR* have interpreted Región as a metaphor for Spain as a whole. Such readings make Región's inaccessibility an allegory for Spain's geographical position as an appendage to the European mainland, for its unique historical and cultural position due to the Moorish conquest and proximity to Africa, and even for single historical events such as

France's closing of its border with Spain in 1946. Spain, in other words, became after World War II Europe's "region" or province, the location of everything backward and atavistic, a place gripped by a time warp where fascism still had not been defeated. In relation to this interpretation, which draws upon the narrative references to the Civil War, the character of Numa, a hunter who kills all those who wish to leave Región, would then symbolize Franco—except that his mythical qualities actually derive from Benet's reading of the "King of the Wood" chapter of James G. Frazer's *Golden Bough*. There, a number of figures similar to Numa are described, including the priests of Nemi and the goddess Diana. Benet's narrator readily identifies the geography of Región as paradigmatic for the rest of Spain during the civil war: "The whole course of the civil war in the Región sector begins to be seen clearly when one understands that, in more than one aspect, it is a paradigm on a lesser scale and with a slower rhythm than peninsular-wide events" (*Return* 63; "Todo el curso de la guerra civil en la comarca de Región empieza a verse claro cuando se comprende que, en más de un aspecto, es un paradigma a escala menor y a un ritmo más lento de los sucesos peninsulares" *VAR* 75). The proletarian uprisings occur as a "effect of mimcry" (63; "efecto de mimetismo"; 75). While in his translation Rabassa chooses mimicry to translate Benet's *mimetismo*, it seems clear that Benet is aiming at the concept of "mimesis" as representation in general—a theme we have been following in the form of the mimesis-diataxis-diegesis triad. Deliberately, however, at different points of the novel Benet presents alternate diataxes that conflict with each other.

On the one hand, the reduction of a long topographical description of the mountains of Region, extending from page 36 to page 42 of the text, would make a case for Región, in its inaccessibility, isolationism, and mountainous terrain, belonging particularly to northern Spain. The citation of a specific parallel of latitude—42° 45"—on page 40 seems conclusive, although also a bit off, as does the altitude of El Monje at 2480 meters on page 36. Región, then, must lie in the mountainous region of western Asturias, the only place in Spain outside of the Pyrenees where peaks of more than 2,000 meters are found. This mountainous topography is entirely different from the hill country of Faulkner and the *sertão* of da Cunha, but it leads in similar fashion to the isolation and backwardness of the inhabitants, as described in a standard "scientific" work such as M. de Terán's and L. Sol Sabaris's *Geografía regional de España*: "The dispersion of its population and the rural character of the settlements become more accentuated towards the interior, as does a way of life and a culture full of archaism ['impregnados de arcaísmo']" (59). Benet's long description of the formation of the Cantabrian range parallels Ferlosio's description of the Jarama. It could have been drawn from an atlas or specialized geological textbook. The emphasis is on the violence of uplift, folding, and collision that formed these mountain ranges: "The Región range shows itself like an enigmatic witness, little-known and disquieting, of all that disorder and those paroxysms" (*Return* 31; "La sierra de Región se presenta como un testigo enigmático, poco conocido e inquietante, de tanto desorden y tanto paroxismo"; *VAR* 39). This description, along the lines suggested by Rivero above, superimposes or condenses geological with political violence.

Other features of Región form a mosaic gathered from various provinces, including Santander, Cáceres, Ciudad Real, Soria, Zaragoza, Teruel, Valencia, Murcia, Jaén, Seville, Málaga, and Almería (see Margenot, "Cartography" 341). In this passage, the narrator inserts Región into a metonymic chain of strongholds of Republicanism by mentioning it "in the same sentence with those of Madrid and Valencia, the two successive capitals of the Second Spanish Republic" (Compitello, *Ordering* 102). Benet thus challenges us openly with the process which Paul de Man has described in his *Allegories of Reading* whereby metaphorical identification functions as a series of metonyms. Like Marcel Proust's novel in de Man's reading, Benet's text "valorizes metaphor [i.e., Región for Spain] as being the 'right' literary figure, but then proceeds to constitute itself by means of the epistemologically incompatible figure of metonymy [i.e., Región as a region of Spain]. The critical discourse reveals the presence of this delusion and affirms it as the irreversible mode of its truth" (18). Metaphorical identification, which seemingly shows us essence directly by superimposing images, in fact works only through a series of displacements. In this way, geographical description becomes Benet's method for exploring the hiatus of discourse between its metaphoric or allegorical mode (where Región represents Spain as a whole) and metonymic or realist mode (where Región becomes just one example of the Civil War). That is, the process of diataxis itself is foregrounded. This diataxis, in turn, challenges traditional Civil War fiction, nearly all of it in the social realist mode, which assumes the exemplarity—achieved through the literary devices of mimesis, empathy, and catharsis—of the individual cases it depicts. (It is worth mentioning that prepublication censorship had ceased in 1966, and the year of *VAR* also saw the first novel published in Spain by a former Republican, Ángel María de Lera's *Las últimas banderas*.) It also questions the relationship between region and nation. In one sense, nations are a larger unity superimposed on a set of regions—Switzerland, with its four national languages, can be viewed this way; in another sense, however, nations elevate a particular region to the status of general representativeness. The Spanish language, for example, is really *castellano*, the dialect of Castile.

Again, *OS* may have served as a model for Benet's cartographic paradoxes. The first sentence of *OS* seeks to construct continuity and completeness in Brazilian topography: "The central plateau of Brazil descends, along the southern coast, in unbroken slopes, high and steep, overlooking the sea; it takes the form of hilly uplands level with the peaks of the coastal mountain ranges that extend from the Rio Grande to Minas. To the north, however, it gradually diminishes in altitude, dropping eastward to the shore in a series of natural terraces which deprive it of its primitive magnitude, throwing it back for a considerable distance in the direction of the interior" (*Rebellion* 3; "O planalto central do Brasil desce, nos litorais do Sul, em escarpas inteiriças, altas e abruptas. Assoberba os mares; e desata-se em chapadões nivelados pelos visos das cordilheiras marítimas, distendidas do Rio Grande a Minas. Mas ao derivar para as terras setentrionais diminui gradualmente de altitude, ao mesmo tempo que descamba para a costa oriental em andares, ou repetidos socalcos, que o despem da primitiva grandeza afastando-o consideravelmente para o interior"; *Os Sertões* 5). In this largest of overviews, everything is accounted for. There are no hidden pockets, and we are able to distinguish relative magnitudes. Once we enter the area of the *sertão*, on the other hand, the map begins to fail: "Our best maps, conveying but scant information, show here an expressive blank, a hiatus, labeled Terra Ignota, a mere scrawl indicating a problematic river or an idealized mountain range" (9; "As nossas melhores cartas, enfeixando informes escassos, lá têm um claro expressivo, um hiato, Terra ignota, em que se aventura o rabisco de um rio problemático ou idealização de uma corda de serras"; 10). A few sentences earlier, just such a problematic river, the Geremoabo segment of the Vasa-Barris, had been described as a "cartographer's fantasy" (8; "fantasia de cartógrafo"; 10). Perhaps it is this "expressive blank" that allows da Cunha to suppress what he knows: that the Vasa-Barris is the river that nourishes and protects Canudos; and that the river will literally run red with blood when the campaign begins. Although da Cunha's scientific scaffolding dictates the ordering of his text into the great blocks of meaning reminiscent of the Brazilian plateau, the author also provides for discontinuities, for the abstention from cause and effect, and for the cultivation of certain "hiatuses" where the political struggle can appear in its natural environment. Da Cunha's mental map of Brazil, whose function is to explain the Canudos campaign, will fulfill that function by pointing in dismayed silence to its own blank spaces, as Benet's points to the disjunction between metaphorical and allegorical modes of reading.

The continuity of mimesis and of diegesis fail at the same time. For example, da Cunha engages in prolepsis by taking as his hygrometer in the climatological section dead soldiers and horses from the campaign (their natural mummification indicates the extreme dryness of the region), described with all the pathos normally reserved for the unfortunate rebels in the second section, entitled "Man" ("O Homem"). The reader is shocked and disturbed to encounter these—as yet unexplained and unjustified—casualties of war in the midst of da Cunha's description of their "natural" setting. No less shockingly, the *terra ignota* theme will be repeated hundreds of pages later in the form of morality, as da Cunha protests by using indirect discourse the inhuman treatment of captured rebels by the government troops: "Canudos was appropriately enough surrounded by a girdle of mountains. It was a parenthesis, a hiatus. It was a vacuum. It did not exist. Once having crossed that cordon of mountains, no one sinned any more" (*Rebellion* 444; "Canudos tinha muito apropriadamente, em roda, uma cercadura de montanhas. Era um parêntese; era um hiato; era um vácuo. Não existia. Transposto aquele cordão de serras, ninguém mais pecava"; *Os Sertões* 382). As Roberto González Echevarría points out, the terms "parêntese" and "hiatus" are from grammar and metrics, respectively: "There is a haunting propriety to this gap being termed a hiatus, as if it were an interruption in the flow of verse, a stop to avoid the cacophony of contiguous vowels with similar sounds, itself a break from some superior law that is about to generate something anomalous. . . . The peculiar expressivity of the blank contains that of nature as represented by the transcendental language of the text, for this is the place where the ultimate mutation takes place. . . . This superior language to which Euclides often appeals is that capacious language—one that can scribble a problematic river on a map—which, like the *sertão*, can absorb it all, even its antagonists, like Conselheiro and his followers" (138).

The reader in the map

The foregrounding of diataxis in Benet and da Cunha, then, stems from a single root: the struggle of language to absorb a landscape. Besides informing the language of the narrator of *VAR* with such "impasses," Benet also gives us a view of their relation to tragic vision, or at least to human passion, by making one of the main characters of his novel a "fantastic geographer." Colonel Gamallo, fighting for the Francoists, conquers eventually the Republican resistance of Región. The narrative reveals that Gamallo wishes to conquer Región as an act of personal revenge for having lost his money and his girlfriend in a casino there. His campaign map is one of pure desire, the toponymic naming reduced to his personal interest in the site in question. The colonel's peculiar mapping can be observed in the description of his campaign headquarters: "One of the rooms . . . had been papered with all the 1:50,000 scale maps of the Torce valley (many of which were simply only blank spaces surrounded by altitude curves of doubtful accuracy), daubed with crosses, rhumb lines, tricky ellipses, and enigmatic inscriptions: 'Pile of files,' 'the dead donkey,' 'here the shepherdess,' 'we return'" (*Return* 56; "Una de las habitaciones . . . había sido empapelada con todos los 50.000 del valle del Torce (muchos de los quales no eran sino áreas en blanco rodeadas de curvas de nivel, de dudosa verosimilitud), pintarrajeados de cruces, rumbos, elipses peludas e inscripciones enigmáticas: 'Montón de fichas', 'el burro muerto,' 'aquí la pastora,' 'volvemos'"; *VAR* 66-67). What diataxis is presented here? Does the mimesis bring forth the symbols? And do the symbols amount to a diegesis, or merely recall one? The "pile of poker chips" refers to Gamallo's gambling loss, while "we return" alludes to his revenge. The combination of blank spaces on the map with the hermetic renaming of landmarks by Gamallo makes of Región a *terra ignota*. But this has been the case since the opening sentence of the novel, as we have seen. Like da Cunha, Gamallo defeats our expectations for continuity—the absence of blank spaces—in a map. Here we see the influence not so much of Faulknerian narrative as of his drawn map of Yoknapatawpha, where toponyms are the events that occurred there. Following the paths of these texts' unexpected discontinuities, we come across unexpected and disturbing continuities as well. For example, Benet frequently provides a continuity of voice between what we had thought to be distinct narrators. On page 208 of *VAR*, the doctor Sebastian, one of the novel's protagonists and narrators, is engaging in geographical description, when he is suddenly interrupted by the book's narrator. The geographical description continues with no loss of continuity. Only the double black trace of the quotation mark tells us that the words are no longer the doctor's. Eight pages later, another set of quotation marks, plus the formula of "dijo el doctor," tells us that the narrative has once again flowed into the doctor's discourse. Nothing else tells us this, for the subject-matter and language of the two voices are indistinguishable. Like Colonel Gamallo's cartography that collapses distance and destination, Benet transgresses the distance between subject and object, between the observer and the observed. This process begins with the title of the novel, which seems to directly address the reader, to involve him in the geography about which he is to read.

The narrator's use of "the traveler" ("el viajero") continues to oscillate between the hypothetical and the real, the pronominal and the nominal, the universal and the particular. For example, sometimes the word "viajero" is capitalized, while at other times it appears in lower-case. In the following sentence, the traveler seems to be hypothetical, to exist as a rhetorical figure for the focusing of geographic description: "The shrewd and persistent traveler, ready to advance a half a mile a day—cutting his way through the underbrush on the lower mountain with a machete"(*Return* 193; "El viajero avispado y tenaz, dispuesto a avanzar a razón de un kilómetro al día—talando a machetazos una broza en el monte bajo"; *VAR* 213). The next sentence begins with "En ocasiones" which gives it a similar hypothetical ring. However, the sentence goes on to suddenly draw the previously abstract and hypothetical traveler into the story by attacking him with the very landscape for which he is the descriptive vehicle: "when the traveler sits down to contemplate the path he has covered and at the moment he lifts his canteen to quench his thirst with a drink of cool wine, a sound that is familiar to him comes to distract his attention immediately" (193; "cuando el viajero toma asiento para contemplar el camino recorrido y en el momento en que levanta su cantimplora para calmar su sed con un trago de vino fresco, un sonido que le es familiar viene a distraer instantáneamente su atención"; 213). This unexpected moment is the beginning of a short ghost story, about the hearing of unexplained noises of a gas motor and the appearance in the night of phosphorescent light. At the climax of this story, the second-person familiar of the title reappears, in a further particularization:

> How wrong you are, my good fellow! Shortly before midnight an unexpected light . . . dazzles him and wakes him up. Confused, obliged to protect his sight with his hands and crawl on his elbows to get out of the underbrush he manages to lift his head. But what are you doing, damn you? Before a circle of phosphorescent light, in the center of which a red light wrapped in heady vapors blinks, disappears into the night with furious wing-flapping and croaking, a terrible and instantaneous barb sinks into his back at the level of his kidneys, knocking him to the ground again with cries of pain and tears of fright. (*Return* 195)

> ¡Cómo te equivocas, paisano! Un poco antes de la medianoche una luz inesperada . . . le deslumbra y despierta. Ofuscado, obligado a protegerse la vista con las manos y a gatear con los codos para salirse de entre las matas, acierta a levantar la cabeza. Pero ¿qué haces, desgraciado? Antes de que un círculo de luz fosforescente, en cuyo centro paradea una luz roja envuelta en capitosos vapores, se desvanesca en la noche entre furiosos aleteos y gra-

> znidos, un terrible e instántaneo aguijón se hunde en su espalda a la altura de sus riñones para derrumbarle de nuevo al sueleo entre gritos de dolor y lágrimas de miedo. (*VAR* 215-16)

This scene dramatizes the descriptions of the unusual fauna of Región given in the first pages of the book, echoes the geological description of the "disorder and paroxysm" out of which Región was created, and foreshadows the deaths with which the novel ends. Proportional to the demotion of the traveler from a model to a particular agent is the increasing confusion of the description itself. The bird's-eye view of the landforms in their larger, epic scope has been reduced to a view of the fauna without coherence: "the landscape, at other times viewed synchronically in its historical verticality as a battleground since ancient times, is now described in its particular configurations, as by one who is himself lost in the brambles; but it becomes paradoxically more indistinct in proportion to the precision with which it is depicted" (Nelson 3). This personalization of the traveler as the center of geographical description derives undoubtedly as well from *OS*, since it occurs in the Brazilian work with great frequency. The birds-eye view of da Cunha's maps fails gradually, replaced by the semi-blindness of the traveler, who becomes more and more real to the extent that he becomes more and more enmeshed in his surroundings, taking his bearings from the imprints left on his flesh:

> The traversing of the backland trails is then more exhausting than that of a barren steppe. In the latter case, the traveler at least has the relief of a broad horizon and free-sweeping plains. The *caatinga*, on the other hand, stifles him; it cuts short his view, strikes him in the face, so to speak, and stuns him, enmeshes him in its spiny woof, and holds out no compensating attractions. It repulses him with its thorns and prickly leaves, its twigs sharp as lances; and it stretches out in front of him, for mile on mile, unchanging in its desolate aspect of leafless trees, of dried and twisted boughs, a turbulent maze of vegetation standing rigidly in space or spreading out sinuously along the ground, representing, as it would seem, the agonized struggles of a tortured writhing flora. (*Rebellion* 30)

> Então, a travessia das veredas sertanejas é mais exaustiva que a de uma estepe nua. Nesta, ao menos, o viajante tem o desafogo de um horizonte largo e a perspectiva das planuras francas. Ao passo que a caatinga o afoga; abrevia-lhe o olhar; agride-o e estonteia-o, enlaçado na trama espinescente e não o atrai; repulsa-o com as folhas urticantes, com o espinho, com os gravetos estalados em lanças; e desdobra-se-lhe na frente léguas e léguas, imutável no aspecto desolado: árvores sem folhas, de galhos estorcidos e secos, revoltos, entrecruzados, apontando rijamente no espaço ou estirando-se felxuosos pelo solo, lembrando um bracejar imenso, de tortura, da flora agonizante. (*OS* 29)

Later, this same strategy will be used to describe the conflict, which had the classic characteristics of a guerilla war. The traveler is now a soldier in one of the expeditions, making his way through the *caatinga*: "Suddenly, from the side, close at hand, a shot rings out" (191; "De repente, pelos seus flancos, estoura, perto, um tiro"; 163).

He is being fired upon from both sides by an invisible enemy. "The bullet whizzes past them, or perhaps one of their number lies stretched on the ground, dead" (191; "A bala passa, rechinante, ou estende, morto, em terra, um homem"; 163). The battle continues for several pages, until the soldiers are completely demoralized by the invisible snipers. And these snipers then emerge from their hiding places, in a foreshortened recapitulation of the effect of topography on the human: "And when the last of the soldiers are out of sight, beyond the roll of the hill there rises up from among the stone heaps—like a sinister caryatid amid these cyclopic ruins—a hard and sunburned face, followed by the rude and leather-clad torso of an athlete. Running swiftly up the steep sides of the ravine, this tragic hunter of armed men is gone in a few moments' time" (194; "E quando as últimas armas desaparecem, ao longe, na última ondulação do solo, desenterra-se de montões de blocos—feito uma cariátide sinistra em ruínas ciclópicas—um rosto bronzeado e duro; depois um torso de atleta, encourado e rude; e transpondo velozmente as ladeiras vivas desaparece, em momentos, o trágico caçador de brigadas"; 165). As man becomes one with his environment in this narration, so too the environment itself becomes humanized as the backdrop to Greek culture. In that descriptive excess which is the hallmark of his style, da Cunha provides mythological allusions in "caryatid, "cyclopic," "athlete," and "tragic." This classicizing tendency is yet another discontinuity in da Cunha's map: it civilizes the barbarians by removing them from the environment which was supposed to explain them. It is as if the narrator, like a drowning man, were casting his allusion at ancient Greece in order to keep from sinking into his own description of the *sertão* as his art is devoured by the landscape.

In contrast, Antonio Conselheiro, the leader of the rebellion, da Cunha's nemesis, blends perfectly into his environment. Conselheiro's insanity is the condensation of the abnormality of Brazil's geographic and racial hiatus, the *sertão*: And so da Cunha uses a geological metaphor to describe him:

> It was natural that the deep-lying layers of our ethnic stratification should have cast up so extraordinary an anticlinal as Antonio Conselheiro. . . . The metaphor is quite correct. Just as the geologist, by estimating the inclination and orientation of the truncated strata of very old formations, is enabled to reconstruct the outlines of a vanished mountain, so the historian, in taking the stature of this man, who in himself is of no worth, will find it of value solely in considering the psychology of the society which produced him. As an isolated case, this is one lost amid a multitude of commonplace neurotics; it could be included under the general category of progressive psychoses. (*Rebellion* 117)

> É natural que estas camadas profundas da nossa estratificação étnica se sublevassem numa anticlinal extraordinária—Antônio Conselheiro. ... Da mesma forma que o geólogo interpretando a inclinação e a orientação dos estratos truncados de antigas formações esboça o perfil de uma montanha extinta, o historiador só pode avaliar a altitude daquele homem, que por si nada valeu, considerando a psicologia da sociedade que o criou. Isolado, ele se perde na turba dos nevróticos vulgares. Pode ser incluído numa modalidade qualquer de psicose progressive. (*OS* 102)

An anticline is an arch of rock uplifted through geologic pressure. Newer strata of rock, exposed to erosive forces, are carried away to reveal the older formations beneath them. Da Cunha seems to seize on this "quite correct metaphor" as a geologic symbol of political and cultural atavism. His insistence on the anticline as a reconstruction on the part of the geologist points to his own authorial role, predicting his deliberate avoidance of a portrayal of Conselheiro's actions during the battle. In a text filled with imaginative depictions (a few of which we have already seen), da Cunha eschews showing Conselheiro during the battle, thereby sowing his text with yet another hiatus. Da Cunha's imagery here anticipates and summarizes his presentation of Conselheiro as a "hollow man," a Brazilian Kurtz, the locus of a collective psychosis. Canudos, too, is an anticline emerging from the flora of the region. In another effort at fantastic geography, da Cunha pauses to name the "pipe reeds, shrubby, hollow-stemmed heliotropes, streaked with white and with flowers that grow in spiked clusters, the latter species being destined to give its name to the most legendary of villages" (33; "canudos-de-pito, heliotrópios arbustivos de caule oco, pintalgado de branco e flores em espigas, destinados a emprestar o nome ao mais lendário dos vilarejos"; 31). Other geological metaphors will make this psychosis a part of Brazilian normalcy. Da Cunha accompanied the fourth military expedition against the insurgents of Canudos (the first three had been defeated), believing that the backlanders represented both a monarchist threat to the young Brazilian republic and an inferior *mestizo* race that would be absorbed eventually into or crushed by the Europeanized coastal Brazilians. The opposition between littoral and *sertao* is essential to da Cunha's geography. Instead of reducing itself to smaller, poorer models of the settlements of the European conquerors, everything moving to the interior of Brazil became larger, more robust, and more difficult of extermination in the name of "civilization." In Note v, for example, the *sertanejo*s are described as the "bedrock of our race" (481; "rocha viva da nossa raça"; 414). The backlander is then compared to the solid granite lying beneath the surface variety of a hill. "The deeper he goes, the closer the observer will come to the definite matrix of the locality in question" (481; "Assim à medida que aprofunda o observador se aproxima da matriz de todo definida, do local"; 414). Just so, as one penetrates into the backlands, the "generalized miscegenation . . . has given rise to every variety of racial crossing; but, as we continue on our way, these shadings tend to disappear, and there is to be seen a greater uniformity of physical and moral characteristics" (481; "a mestiçagem generalizada produz, entretanto, ainda todas as variedades das dosagens díspares do cruzamento. Mas à medida que prosseguimos estas últimas se atenuam. Vai-se notando maior uniformidade de caracteres físicos e morais"; 415). From its central paradox of locating a place by going deeper and deeper, by carrying away the mountain in order to excavate its truth, the allegory goes on to produce at least three further aporias: 1) the farther one descends into miscegenation, the more stable race becomes; 2) looking back to the original description of the backlands, Brazil's identifiable location is *Terra ignota*; 3) the bedrock of Brazilian society is a weak race that will be crushed by the stronger. Brazil's "rocha viva" (literally, "live rock") is moribund.

As the campaign dragged on, the social Darwinist da Cunha came to realize that the superior evolutionary forces could win only through an atavistic surrender to the forces of the *sertão* itself, and this process became the kernel of various narratives within his text. One example is a description of the end of the second Canudos campaign, when the soldiers had to retreat partly because of starvation. At the end of their last battle, a herd of wild goats invades the battlefield and the soldiers slaughter them: "And an hour later these unhappy heroes, ragged, filthy, repulsive-looking, could be seen squatting about their bonfires, tearing the half-cooked flesh as the flickering light from the coals glowed on their faces, like a band of famished cannibals at a barbarous repast" (*Rebellion* 223; "e, uma hora depois acocorados em torno da fogeiras, dilacerando carnes apenas sapecadas—andrajosos, imundos, repugnantes—agrupavam-se, tintos pelos clarões dos braseiros, os heróis infelizes, como um bando de canibais famulentos em repasto bárbaro"; *Os Sertões* 194). The superfluous and inexact use of "cannibals" indicates da Cunha's relish for indicating the loss of civilization experienced by the traversers of the *sertão*. It also echoes parodically the repeated labeling by press and government of the rebels as "cannibals." The fourth campaign undergoes a similar reduction: as the government troops take over the houses of the *sertanejos* and throw off their uniforms for attire better suited to the heat, "with a little presence of mind, a *sertanejo*, making his way through an opening in that extensive enclosure, would be able to take his place among our soldiers, with his rifle over the top of the barricade" (454-45; "com alguma presença de espírito, o sertanejo pudesse insinuar-se pelos rombos do tapume extenso, e ali se quedar forrando-se às torturas do erco, sem que o conhecessem"; 390). This description anticipates the final sentences of his work, when the civilized victors twice exhume the corpse of the rebel leader, decapitate it, and send the head as a trophy to the coast, "where it was greeted by delirious multitudes with carnival joy. Let science here have the last word. Standing out in bold relief from all the significant circumvolutions were he essential outlines of crime and madness. The trouble is that we do not have today a Maudsley for acts of madness and crimes on the part of nations" (476; "onde delivaram multidões em festa. . . . Que a ciência dissesse a última palavra. Ali estavam, no relevo de circunvoluções expressivas, as linhas essenciais do crime e da loucura. É que ainda não existe um Maudsley para as loucuras e os crimes das nacionalidades"; 408-09; Da Cunha alludes here to Henry Maudsley, a noted psychologist who specialized in the identification of criminal mentalities). And so, even the insane millenarianism of Conselheiro has proven true: "the backlands will turn into seacoast and the seacoast into backlands" (117; "então o certão virará praia e a praia virará certão"; 115). That is, the barbarians will become civilized, and the civilized will become barbarians.

Passages such as these, which bind together terrain, the human, and fate, find echoes in Benet's invention of "characters" such as the hunter, Numa. The name itself alludes to Numa Pompilius, a legendary king of Rome. Numa, like Antônio Conselheiro, can be seen as the anticline of Región's geography, both physical and cultural. As the assassin of any stranger who tries to enter the forest, he represents the principles of isolation and incomprehension that keep Región "regional" rather than central. He emerges suddenly from the topography like the sniping backlander of *OS*, and is in fact a sniper: the novel's last sentence reports the "echo of a distant shot [which] came to re-establish the habitual silence of the place" (*Return* 315; "el eco de un disparo lejano [que] vino a restablecer el silencio habitual del lugar"; *VAR* 125). Most readers can imagine that this shot has killed Marré Gamallo, daughter of the colonel, who has come to Region to question Doctor Sebastián about her past. Knowing that Benet's father was killed in the war, we might also think of this shot as a symbolic repetition-compulsion. Our conclusion must be that the desire to understand oneself will always be defeated. Within *VAR*, Numa is explained alternatively as an old Carlist soldier who never surrendered, as a renegade monk, and as a disappointed lover seeking his revenge like Colonel Gamallo (251). Ultimately, however, such explanations seem as fallible as the characters who utter them, and as ready for defeat at the hands of inaccessibility as the traveler we have pitied above. Numa represents symbolically Región's forbidding geography in all its aspects: "A warm breeze blew like the senile breath of that old and woolly Numa, armed with a carbine, who hereafter will guard the forest, keeping watch night and day over the whole extent of the estate, firing with ineffable aim every time footsteps in the leaves or the sighs of a weary soul disturb the tranquility of the place" (6; "Sopló un aire caliente como el aliento senil de aquel viejo y lanudo Numa, armado de una carabina, que en lo sucesivo guardará el bosque, velando noche y día por toda la extensión de la finca, disparando con infalible puntería cada vez que unos pasos en la hojarasca o los suspiros de un alma cansada, turben la tranquilidad del lugar"; 12).

Both *VAR* and *OS*, to take up Jo Labanyi's point, make the reader into a traveler: "The text is 'about' its subject-matter not in the sense that it contains it but in the sense that it wanders round it as an absent centre. . . . It provides us not with a gradual recovery of past events but with a series of digressions 'around' them" (132-33). Da Cunha's narrative becomes strained by the paradoxes of his position, resulting in an involved syntax, sesquipedalian vocabulary, and a continual shifting in narrative position and rhetorical strategy. If, as Denis Cosgrove claims, landscape serves ideology, then the landscapes created by these texts serve "ideological difference." Although on the surface they are of different genres, each is troubled by the discourse of the Other—tragedy and novel color the "science" of *OS*, while the discourses of geology and engineering take over the diegesis of *VAR*. Both Benet and da Cunha produce imaginary discourses where the processes of immersion in and active identification with the natural environment negate the possibility of the birds-eye view, of extracting oneself from that environment in order to a coherent mental map. The failure of both these geographers to "control the imaginary" parallels the problems of national political control that form the theme of their narratives.

Chapter Five

Solipsistic Regions, *Fogo morto*, and *The Sound and the Fury*

In this chapter I begin with two interludes based on my own exploratory experience of the regions written about by Faulkner and Lins do Rego, in an attempt to ground the similarities of their regionalisms in the particularities of geography.

Figure 10. Rural road in Lafayette/Yoknapatawpha County, Mississippi

Day one

From Oxford to Jefferson and Lafayette to Yoknapatawpha: it is well known that the hotel Faulkner refers to is the Peabody, in the heart of downtown Memphis. No need to check that out, then. Instead, I point my rental car south and head for Clarksdale, Mississippi. My plan is to approach Yoknapatawpha from the West across the Delta, to see the hill country "stand up" like the first explorers did, following in the footsteps of the medical student Garrett Elliott Pendergrast in 1802, when the easy route into the Mississippi Territory was by river, and then eastward, rather than overland through thick forest. The Territory was still raw then, populated more by French and Choctaws than by English speakers. Its only urban places were Natchez, New Orléans, and Mobile. Pendergrast, whose main interests lay in observing the flora and the diseases of the region, described the soil below Memphis: "The country . . . on the east side of the river, extends as far as the upper branches of the Tombigby, pretty uniformly composed of a tough clayey soil, interspersed, occasionally, with strips of good land, on the banks of some of the larger streams; and free from its morasses in its whole extent." (13). I think about Edouard Glissant driving these same routes. Teaching for a year or a semester at Southern University in Baton Rouge, Louisiana, Edouard Glissant and his friends drove up to Oxford to visit Rowan Oak. They had previously visited the Museum of Rural Life in Baton Rouge, where the objects, tools, and atmosphere reminded Glissant of his West Indian childhood. Rowan Oak presented yet another diataxis, in between the other two: "I did not feel the absence (the eviction, the erasure) of Blacks as I had at Nottoway [Plantation]. Nor did I feel their presence. . . . It was as though the aura of [Faulkner's] works had elevated the building and its surroundings to a state of splendid indifference, so that they transcended their origins. Can literature make one forget grief and injustice? Or, rather, is literature, and particularly the work of Faulkner, inextricably tied to grief and injustice so as to be able to point them out or fight against them?" (14). I am fascinated by this Caribbean writer's discovery in Faulkner of the same overpowering sense of geography and landscape, of microclime, as I have. For a writer like Glissant, who believed that location and history held more importance for identity than disembodied notions of race, Faulkner could serve as a *pierre de touche*. Fellow Caribbeans. The famous first sentence of William Faulkner's essay on his home state provides a good example of a mental map. It unites disparate elements of topography (the lobby of a hotel room and the Gulf of Mexico); and the well-known paradox of Mississippi beginning in Tennessee raises the question of how, and by what criteria, geographical and political boundaries get drawn. One solution to the immediate paradox is to realize that Faulkner's Mississippi is not a state, but an imagined community. In Faulkner's work, we are reminded, Yoknapatawpha—the true imaginary county where most of his best work takes place—stands for Mississippi, Mississippi for the South, the South for the US, and the US for all the affairs of the human heart.

The diataxis of the "Mississippi" essay moves jarringly back and forth between narration and description (diegesis and mimesis), often combining the two, for example, when Faulkner recounts his exploits as a smuggler off Mississippi's southern coast. Narrative in the essay comes in several flavors: historical (the Civil War or great floods of the Mississippi); autobiographical (Faulkner refers to himself in the third person as "the boy," "the young man," "the middle-aged novelist," etc.); and fictional (Jefferson, the county seat of Yoknapatawpha, is referred to, but not Oxford, and Snopes, the white-trash family that eventually takes over Jefferson, are everywhere). Just as da Cunha chose the "establishing shot" for the opening of *Os*

Sertões, Faulkner begins his essay on a grand scale, encompassing the whole latitudinal plunge of Mississippi (longitude follows). He ends it in the parlor of Rowan Oak (his estate), as his servant Mammy Barr declines toward her end, and "the middleaging" ("Mississippi" 42) delivers the sermon at her funeral. One device of the diataxis is the refrain, "loving it and hating it," which serves as a switch point from narrative to description or harangue, and is reminiscent of Shreve's "Why do you hate the South?" jibe at the close *of Absalom! Absalom!* The swooping and diving, amplification and stretto of the essay seem designed to make their own theoretical point about geography: landscape is a creation of narrative, and vice versa.

Day two

But as I drive the next morning (after a side-trip into Arkansas to verify that there is a West Helena, but no East Monroe) from Clarksdale across the seemingly endless Delta towards Lafayette County (the real-life model for Yoknapatawpha), I see that reality is more complicated. The Delta is completely flat, the earth black, the evidence of cotton-growing everywhere. The lack of topographical features in the Delta made the organization of large-scale plantations easier, almost inevitable. I realize that the Delta could never have become the topography of Faulknerian fiction—there are no microclimes, no places to keep secrets. Or, conversely, had Faulkner grown up here, his fiction would have been vastly different, or he would have found another line of work. The movement back and forth between scenes from the Delta and from the hill country, without legend or explanation, makes problematic the photographic collection of Alain Desverges. The Delta was not Faulkner's any more than one would travel to Jefferson in order to imbibe blues music. Elmo Howell notes that had Faulkner written about the Delta, he would have been Stark Young: "Stark Young, like Faulkner, writes from his own experience in Mississippi, but the Como and Sardis Country in Panola County conforms more closely to the Delta pattern, and Young's novels become stifling with the elegant sameness of plantation life. Faulkner was more fortunate in his geography" (77).

As I cross the line into Panola County, I catch a first glimpse of low hills in the distance. Soon the houses on either side of the highway are built on little rises. As I enter eastern Lafayette County, the hills become more pronounced; the earth has lost its blackness and is now brown or at times red, clayey, rather than black. Cotton has disappeared in favor of cattle and horses. The possibilities for extensive plantations have diminished in this hill country, with its many accidents of landscape. Sutpen's Hundred is an anomaly, a microclime, multiplying the opportunities for the secrets Glissant talks of in the epigraph. This is Faulkner country. He was born in New Albany in Union County in 1897; his grandfather, the Old Colonel, whose death is recounted in "An Odor of Verbena," was the mover and shaker of Ripley, in Tippah County. I approach Oxford through New Albany and Ripley, stopping there to see with my own eyes that they belong to the hill country and that someone who awoke suddenly in any one of them might not be able to say exactly where he was. Some have posited that the geography of Yoknapatawpha is derived from Tippah and

Union counties, rather than from Lafayette, even if Jefferson is without doubt a truly imagined and subtly transposed Oxford (see McHaney 263; Luce 77). Mottstown, for example, featured in *As I Lay Dying* and *The Sound and the Fury*, with its crossing railroad lines, seems modeled on New Albany. But the roads that Faulkner drew for his fictional county match perfectly those of Lafayette. All I have to do in order to locate Faulkner's true imaginary places—so I tell myself—is to drive those roads.

In New Albany I comb the Union County Historical and Heritage Museum for Faulkneriana, and the most valuable thing I find for my purpose is a picture of an old barn in a field, painted by William Dunlap as the commemorative for the William Faulkner Centennial Celebration. Dunlap titled his painting "Postage Stamp of Native Soil" after Faulkner's characterization of the focus of his writings on Yoknapatawpha. The painting's relationship to Faulkner is oblique, depicting as it does neither an action nor a scene from any of the Yoknapatawpha novels. The artist wrote of this work: "Rather than invoke some antebellum sentimentality about Mr. Faulkner and Mississippi, I find I've always been more interested in the subtle beauty of the commonplace (as was Mr. Faulkner). . . . I don't think anyone who's ever driven the roads of northeast Mississippi just as an afternoon storm is clearing and that wonderful light pours in from the west can mistake this work for any place but our own" (text with the painting at the Union County Historical Society and Heritage Museum, New Albany).

Faulkner's prose shows that he had a precise interest in the geography of Mississippi: Jones County, hill country, is populated by Irish and Scottish immigrants, hardscrabble farmers who never owned slaves and who are not only racist, but also xenophobic: "No Negro ever let darkness catch him in Sullivan's Hollow. In fact, there were few Negroes in this country at all: a narrow strip of which extended up into the young man's [Faulkner's] own section" ("Mississippi" 33). He remarks, for example, that to travel thirty miles east one has to travel "ninety miles in three different directions on three different railroads" ("Mississippi" 11). The route and the mileage are accurate, like the "12 mi" Faulkner wrote on his map to indicate the distance to Sutpen's Hundred. As Calvin S. Brown has pointed out, Faulkner is extremely accurate in his mileage, directions, and landscape markers. In terms of placing houses in Jefferson, on the other hand, he tends to invent everything wholesale rather than use existing structures. In the evening, on my way to the cemetery and Faulkner's grave, I meet Mr. Leslie, whose father was mayor of Oxford for several decades. We talk about this and that, he shows me his falcon (he, too, is a falkner, that is, a falconer), and I explain why I am here in Oxford, chasing after places. He tells me that old-timers can readily identify place after place in Faulkner's fiction, such as College Hill for Sutpen's Hundred.

Day three

In the morning, after breakfast downtown in a bakery filled with Ole Miss students and faculty, and after adding my own photographic endeavors at the courthouse and the confederate soldier, I drive out towards Sutpen's Hundred. I drive past "the church Sutpen rode fast to" (in fact, the church Faulkner was married in). I keep my eye on the car's odometer. Faulkner wrote that Sutpen's Hundred was twelve miles from Jefferson. At ten miles I cannot go any further: I'm up against the recreation area of Lake Sardis, formed by damming up the Tallahatchie River. I park and walk back up the paved road a bit, then take a narrow dirt road that heads west. It twists and turns for about a mile, until an expanse of bottom land opens away from me. I won't see this much flat land in Lafayette County again. The mud tells me that it is flooded every year by the nearby Tallahatchie. Here, then, exactly where Faulkner marked it on his map (well, it's a hundred square miles, so he had some fudge room), is Sutpen's Hundred (see Figure 11).

I wish to "find" Sutpen's Hundred, precisely because it is one of Faulkner's true imaginary places. If I can just locate the true part, then the imaginary will be enhanced, or so I believe. As pointed out earlier, large-scale plantations were characteristic of the Delta, not of the hill country. Faulkner's imagination located a postage-stamp of soil near Oxford that could make the character of Sutpen feasible.

Figure 11. "Sutpen's Hundred" 10 miles NW of Oxford, Mississippi

Agriscapes

It may be true, as Shelby Foote claims, that "there simply aren't any aristocrats in Mississippi and probably never were any, and Faulkner realized that by the time he had written three novels" (41). Indeed, there are no aristocrats, because Faulkner's world is a frontier, in the words of Glissant,

> not only because the Mississippi River is the invigorating torrent, and the Yoknapatawpha and Tallahatchie rivers, rather than tributaries of branches, are its mythic daughters. Not only because the whole South, and by extension the state of Mississippi and consequently its projection, Yoknapatawpha County, are actually frontier sites; but also and especially because the writing Faulkner has used to re-create these places—this Place—has also literally stirred up something: movement, hesitation, transition, uncertain identities, and truths that cannot escape the charm of the possible and the impossible all mixed together. (227-28)

Out of this telluric situation, however, arise enormous differences between the individual dirt farmer of the hill country and the plantation owners of the bottom land, who saw each other with mutual distrust and disaffection. Geography becomes destiny, and perhaps we can attribute the notorious difficulties of Faulknerian chronology, at the diegetic level, to the clear order he achieves at the mimetic level. This idea of what we might call "geographical fate," which runs through the essay "Mississippi," can be encountered in the fiction as well. Faulkner addresses this fate directly, for example, in this aside to the reader in *The Hamlet*: "Geography: that paucity of invention, that fatuous faith in distance of man, who can invent no better means than geography for escaping himself; himself of all, to whom, so he believed he believed, geography had never been merely something to walk upon but was the very medium which the fetterless to- and fro-going required to breathe in" (215). This passage is meant to explain the story of Jack Houston, who leaves Mississippi heading west at sixteen, only to return and marry seven years later. We can read here Faulkner's own escapes from Mississippi, West and North, to Hollywood and the Royal Air Force, and its inevitable pull on him until he came back.

Later, in the library of Ole Miss I find a soil map of Mississippi (Logan), and discover that Lafayette County is divided between two soil types: brown loam and loess in the west and shortleaf pine in the east. This distinction divided settlements between valley regions covered with fertile loess (e.g., Sutpen's Hundred in the northwest corner), and piney hill country (e.g., Snopes and Bundren country):

> The agricultural settlement patterns of Lafayette County evolved against the environmental backdrop. Land speculators and initial settlers perceived the northern and western parts of the county as having the greatest potential, and these areas were quickly patented. Northern and western Lafayette Country came to have the greatest number of plantations, the largest black populations, and the highest cotton acreages and per acre yields. The area south and southeast of Hilgard's line [between "bottom" and "hill" topography] became primarily the domain of white farmers with small holdings. These basic cultural differences within the country persisted well into the twentieth century. (Aiken 5)

Hence, Faulkner had every reason to locate Sutpen's Hundred of *Absalom, Absalom!* to the northwest of Jefferson, in the proximity of the Tallahatchie and its brown loam and loess, and the hardscrabble Bundren farm of *As I Lay Dying*, the small farms of the negrophobic farmers in the piney hills and tannic soil of Jones County to the southeast, and the Snopes family that destroys with its capitalist greed. Faulknerian diataxis transforms the dividing line of soil types into the dynamics of social interaction and conflict (and in the middle, in the vortex between these two opposing forces, the town of Jefferson, and Compson's Mile, the tract of land that eventually shrinks to the claustrophobic manor house of Benjy, Jason, and their mother. The "Compson House" of *The Sound and the Fury*, with its imposing columns, is in fact right in the middle of Oxford. The plantation miniaturized). Faulkner recurs to this distinction in *The Hamlet*. The narrator has interceded in the journey of Ike Snopes as he goes to rescue Houston's cow from a fire. Snopes runs from Frenchman's Bend in the southwest further south to the hill farm, unaware of the geographic differentiation he encounters: "A mile back he had left the rich, broad, flat river-bottom country and entered the hills—a region which topographically was the final blue and dying echo of the Appalachian Mountains. Chickasaw Indians had owned it, but after the Indians it had been cleared where possible for cultivation, and after the Civil War, forgotten save by small peripatetic sawmills which had vanished too now, their sites marked only by the mounds of rotting sawdust which were not only their gravestones but the monuments of a people's heedless greed. Now it was a region of scrubby second-growth pine and oak among which dogwood bloomed until it too was cut to make cotton spindles, and old fields where not even a trace of furrow showed any more" (174). Through geographical description Faulkner delineates a social difference between landed gentry and dirt farmer.

Similarly, in *Sartoris* Faulkner uses the description of one house in Jefferson in order to distinguish between hill and valley cultures:

> The lot had been the site of a fine old colonial house which stood among magnolias and oaks and flowering shrubs. But the house had burned, and some of the trees had been felled to make room for an architectural garbling so imposingly terrific as to possess a kind of majesty. It was a monument to the frugality (and the mausoleum of the social aspirations of his women) of a hill-man who had moved in from a small settlement called Frenchman's Bend and who, as Miss Jenny Du Pre put it, had built the handsomest house in Frenchman's Bend on the most beautiful lot in Jefferson. The hill-man had stuck it out for two years, during which his women-folk sat on the veranda all morning in lace-trimmed "boudoir caps" and spent the afternoons in colored silk, riding about town in a rubber-tired surrey; then the hill-man sold his house to a newcomer to the town and took his women back to the country and doubtless set them to work again. (44-45)

Day four

I devote my last day in Lafayette County to a trip to this southeastern corner. Varner's Corner comes into being where a north-south highway that crosses the Yocona meets an east-west axis. Today, of course, Varner's store would be a gas station/convenience store, and where I find one at the appropriate intersection, I immediately log it into my mental map of Yoknapatawpha (see Figure 12).

I cross the Yocona (whose name is given in older maps as the Yoknapatawpha) heading south, and turn onto dirt roads as soon as I can. Immediately noticeable is the absence of settlements and farms. From time to time, a thin strip of bottom land appears between the hills that have been cleared for farming. Otherwise, I see only pine trees and gullies. It has rained. The thick layer of red mud and clay on the car makes me fear that I will be kicked out of the Emerald Club when I return it. When I finally come upon a vista that I feel approaches what Ike Snopes saw on his barefoot walk to Varner's Corner, I snap a picture of the Bundren/Snopes environment in all its desolation (see Figure 13).

Faulkner's preoccupation with the geographic implications of life forms a constant in his narrative fiction. As Michael Millgate has noted, Faulkner almost from the first recognized the potential advantage of regionalism, in "exploiting

Figure 12. "Varner's Corner" to the SW of Oxford, Mississippi

[Yoknapatawpha's] fictiveness, in asserting those narrative and symbolic freedoms special-worldliness ideally bestows" (79). As I drive the length and breadth of Lafayette County, I realize how medium-sized it is, about the size of Centre County, Pennsylvania, where I live. I imagine myself knowing Centre Country well enough to narrativize virtually every inch of its topography, the way Faulkner did Lafayette. I am struck by the absolute audaciousness of taking the map of this medium-sized county of average historical and topographic interest, and of "stretching" it by filling it with so many incidents and characters that it fits an entire nation.

Figure 13. "Bundren Country" in the southwest corner of Lafayette County

Regionation

So, too, and even more instinctively and tenaciously, did José Lins do Rego do with the area around Pilar, Paraíba, as we will see in the continuation of this interlude. Both authors were part of an international, interbellum propensity towards rural fiction, a "literature of the soil." For Edward Casey, such a move in literature parallels a return in Western intellectual history "back into place." Historically, it reflects the profound crisis in national governance that eventually resulted in fascism and dictatorship in many places. In the Americas this literary trend was called "regionalismo," or "regionalism": "Regionalism sets itself the task of examining an area, of analyzing it, and of seeing it in its context. Faulkner and do Rego both deal centrally with a region in transition from patriarchal forms to a new social and economic structure. Each uses a long historical view to show how man conquers the land, creates customs and traditions, until a community exists" (Standley 73). Their work carries out in the realm of literature what can happen in reality: "What was formerly a region may through mutational alchemy become a new nation" (Kaplan 44). These authors' "new-old nations" or "true imaginary nations" are heterotopias, alternative imagined communities, constructed in opposition to the dominant political discourse and economic tendencies of Brazil and the United States. Faulkner's fiction in particular is haunted by ideas of defeat in war, of occupation by a "foreign" power, and of the South as a "near-nation"—all this despite or because of the fact that he did not write "Civil War novels." Lins do Rego's work avoids any depiction of the Brazilian Estado Novo (1937-45) under strongman Getúlio Vargas.

Faulkner's most important early works emerge from a context of this regional writing, which exploded starting in 1922: Hubart A. Shands's *White and Black*; T.S. Stribling's *Birthright*; and Clement Wood's *Nigger* were all published in that year. The decade that ended with the appearance of *The Sound and the Fury* (1929) saw one Southern region after another enshrined in a work of fiction, which often centered on African American themes and traditions. Writers tended to specialize in particular areas: Jean Toomer's *Cane* (1923) described events in rural Georgia; Dubose Heyward set his novels, *Porgy* (1925) and *Mamba's Daughters* (1929) in Charleston; Ambrose E. Gonzales had made the South Carolina Sea Islands the scene of *With Aesop along the Black Border* (1923), as did Julia Peterkin in *Scarlet Sister Mary* (1928). As Thadious Davis puts it, "at this point in time the Negro was the South's metaphor" (27). Clearly, it was time for Mississippi to enter the list. In 1929 Faulkner's own *Sound and the Fury* was to follow this trend of using the issue of Black humanity to delineate the boundaries of Southern culture, and the issue of Southern culture to repudiate the millennial inheritance of the United States, in terms best articulated by Barbara Ladd: "When a nation envisions itself in ahistorical and millennialist terms, as new and as redemptive, it denies its relationships to the past, even to the history of its making. For a white southerner who had been prior to his defeat by the United States, at the very center of a southern nationalism that envisioned itself in terms remarkably similar to those of the United States (i.e., as redemptive), the consequence is that . . . he has become through his defeat the inheritor of history and the bearer of prior displacements in his own" (Ladd 153). The regionalist fiction of Faulkner maps the sites of such displacement.

Displacement shows itself, for example, in the disjointed contours of the decadent plantation. *Scarlet Sister Mary* is set in the Quarters of the Blue Brook Plantation, still dominated by the mansion, now uninhabited: "A great empty Big House, once the proud home of the plantation masters, is now an old crumbling shell with broken chimneys and a rotting roof" (Peterkin 12-13). By the same token, Blacks still inhabit the "Quarters," the lack of roads and bridges keeping them isolated and preventing their migration to urban centers. It is not far at all from Peterkin's image to the scene Faulkner set in *Sanctuary* for the rape of Temple Drake: "The house was a gutted ruin rising gaunt and stark out of a grove of unpruned cedar trees. It was a landmark, known as the Old Frenchman place, built before the Civil War; a plantation house set in the middle of a tract of land; of cotton fields and gardens and lawns long since gone back to jungle, which the people of the neighborhood had been pulling down piecemeal for firewood for fifty years or digging with secret and sporadic optimism for the gold which the builder was reputed to have buried somewhere about the place when Grant came through the county on his Vicksburg campaign" (4). Naturally, this hulk is inhabited by a motley crew: Popeye, the gangster; his moll, Ruby Lamarr; Lee Goodwin, the moonshiner; and various locals, identified by their thick accents, overalls, and bare feet. The former mansion now resembles a transient hotel, where pass the likes of Horace Benbow, Temple Drake, and Gowan Stevens. In a sense, Faulkner's fiction is a journey through various ruined mansions: to the Old Frenchman place we can add the Jefferson house inhabited by Dilsey, Jason, and Miss Quentin in the *Sound and the Fury*; the McCaslin house in *Go Down, Moses*; and the short-lived, pretentious manor that crowns Thomas Sutpen's ambitions in *Absalom, Absalom!* Faulkner's knowable community, however, includes rude cabins and modest homes as well, such as that of Lucas Beauchamp, which gives the title to "The Fire and the Hearth."

Scholars of literature have tended to analyze Faulkner's geographies into component parts, race being one of the most frequently visited categories, gender another. Even the material influences on Faulkner's fiction have been split, for example, between nature and built environment (see Kartiganer and Abadie, *Faulkner and the Natural World* and Hines, *William Faulkner and the Tangible Past*). Critics who attempt to bring the various categories of analysis together, as Margie Burns does, nevertheless still formulate them as distinct units, different poker chips that can be traded one against the other: "one function of race privilege, reflected with staggering honesty in Faulkner's narrative[s], is to triangulate class privilege and gender privilege" (112). But in the here and now, walking the streets of Oxford and driving the reivers' road, there seem to be even more factors involved in the construction of the highs and troughs of mental maps. Even more than on my trip to Canudos, I am following in other's footsteps: my report will not be a Stanley Livingston trip through Darkest Mississippi. Indeed, Faulkner's geography has been so frequently retraced that I am afraid my photographic quest may prove superfluous. Way back in 1948, Bradley Smith published several photos of the region, of Rowan Oak, the Oxford Courthouse, mules pulling a wagon, and a "Spanish moss-framed cabin ... standing, decayed and desolate, in the marshy woods . . . typical of Faulkner's South" (86). Willard Pate made his own "Pilgrimage to Yoknapatawpha." The black-and-white photographs of Martin J. Dain's *Faulkner's County: Yoknapatawpha* were all taken in Lafayette County or its environs. The first thirty-three pages of the book describe Oxford, county seat and model for Faulkner's Jefferson. It then moves to Parchman Farm, a penitentiary, then to a liquor store just across the county line, and back to Jefferson and some Oxford personalities, including the mayor and Faulkner's brother. The next thirteen pages deal with churches, both black and white, and the next ten with schools, both black and white. We now move into forms of work and commerce, including an antiquated blacksmith's shop, a farm animal auction, and an itinerant salesman, who allows Dain to wryly remark that "There is no longer an itinerant sewing-machine salesman." Images of desolate, ruined cabins or mansions—images that remind me of the blank waters of Canudos, and the eerie silence surrounding the Engenho Corredor in Pilar—are the preferred mode of capturing the reality Faulkner sought to portray. They tend to freeze time and to give a conservative view of a static society. They should be supplemented with less homogenous, dynamic images that reflect the simultaneity of preservation and change in this area.

John Lawrence and Dan Hise have paid tribute in *Faulkner's Rowan Oak* to the author's home in Jefferson. He bought the mansion and property in 1930, just after he had definitively "moved" his fictional world to his home state. Again, the photographs are in black and white. Color would not be fictional or historical—the *Wizard of Oz* effect. Color could never convey the Compson or Sutpen houses. "The interiors resonate with Faulkner's powerful presence, but they do not charm the eye," say the authors (23). What can be the basis for such a statement? Who resonates in what, or what in whom, and who can tell the beginning and ending? Faulkner's presence makes the house "powerfully ordinary." For one thing, there is the sparseness and lack of obvious creature comforts such as air conditioning, a result not just of the near penury of the serious artist, but also of "Faulkner's insistence that he must not live too comfortably" (23), that luxuriousness would soften the edge that he needed as a writer.

In visiting Rowan Oak, I am impressed by the large-scale version of the map in the hall. It is Faulkner's map of Yoknapatawpha, but with many more events drawn onto it in handwriting that clearly differs from Faulkner's own. Mack Harris Scott embraced Faulkner's geographic impulses, and drew maps of nearly every significant event in Yoknapatawpha and in Jefferson as the essence of his Master's thesis at the University of Mississippi. Scott notes that he "grew up in one of the small towns [of West Tennessee] and this fortuitous circumstance has inevitably colored [his] imagination" (171). Scott's seemingly arbitrary overlaying of West Tennessee onto Central Mississippi to create a third, true imaginary place, testifies to Faulkner's success in making Jefferson a touchstone for much larger group identifications. Only

Scott's vibrantly colored (dare I say, obsessed?) imagination would risk trying to improve on one of Faulkner's own maps—one for *Absalom, Absalom!* in 1936, and another in 1945 for the *Portable Faulkner*—with additional places and events that had shaped Yoknapatawpha County in the intervening years. Or would dare to draw a map of Jefferson, the county seat and scene of *The Sound and the Fury* and parts of many other novels, filling every block with an office or a home. A glance at Scott's map of Jefferson, where line is invaded by text, mimesis by diegesis as each building is endowed with a unique Faulknerian event, shows immediately the immense importance given to the Compson home and property, which sits across from the courthouse and opera house. Significantly, Scott titled his thesis "All Our Yesterdays," a quote from the same speech of Macbeth Faulkner had used as the title for *Sound and the Fury*.

While idiosyncratic, Scott's mapping of the "more secret, fictional recesses" (Durand n.p.) of Yoknapatawpha certainly pays tribute to the ability of Faulkner's fiction to capture the imagination. While Faulkner's map left plenty of empty space that could fool a viewer into thinking of it as cartography rather than narrative, Scott leaves us no doubt, in James Justus's words, that "Faulkner's most familiar technique in his depiction of place—[was] attributing the specialness of particular sites to what happened there. But what about the prior narratives themselves, the referents without which the map rubrics would remain arcane? Following the Faulknerian formula—'people first, where second'—we think customarily of the author's extensive gallery of characters as the propelling energy in the works, that a Snopes or a Compson make their places significant because of who they are and what they do" (23). The Faulknerian map, with its imprimatur of "sole owner and proprietor" points to the Faulknerian mental map as laying out an *a priori* aesthetic according to which history, event, and character must arrange themselves: "Faulkner probably realized that he was fortunate in his geography, but being so was clearly not satisfying to his ambition. If the distinctive quadrants of his maps, each devised as appropriate setting for distinctive human acts, answered to the demands of artifice, so too did those places in his narratives. Like the map-making, Faulkner's style is the assertion of the creator, the assertion of his being, his identity—not those of his characters" (26). In other words, for Justus, diegesis creates mimesis in Faulknerian fiction. Yet, driving around Lafayette County, and perhaps under the influence of Glissant's analysis, I am more impressed by the passion with which Faulkner absorbed the mimesis of local landscapes, agriscapes, and ethnoscapes and turned this into diegesis. The process is in fact dialectic and circular.

One imaginary aspect of a true imaginary place is the almost inevitable feeling of stasis it invokes, like a photograph that denies motion or change. Indeed, Faulkner's fiction shares with Dains's black-and-white photographs the function of causing us to "remember" something no longer necessarily "true," if it ever were true: "Today, people who visit Oxford and Lafayette County to experience the flavor of Jefferson and Yoknapatawpha County are probably disappointed. During the past twenty-five years sweeping social, economic, and spatial changes have come to Mississippi and the entire South. Much of the basic geographical framework of the Lafayette County that Faulkner knew, including some roads and place names, remains, but almost all of the color of his time is gone. Forests dominate the landscape that as recently as 1950 was pasture and cropland. Cotton remains important, but machines have replaced the legions of mules and humans that once toiled in the fields from sunup to sundown from April through October. Oxford is now a progressive small city of 12,000, not the agricultural town that Faulkner knew" (Aiken 19). These words were written in 1977, and are even truer today. I begin to feel guilty for writing on Faulkner, for perpetrating this myth of the timeless South. It may simply be my overvivid imagination that allows me to see a crossroads gas station as the modern incarnation of Varner's store, or perhaps I am merely imitating Edouard Glissant, but I am not disappointed in the least. As I drive around, I come to feel that those who are disappointed have not read Faulkner carefully, because his work is from first to last about historical change, and about the ability or inability of individuals to adapt to it. That is indeed a major theme *of The Sound and the Fury*, to be analyzed in chapter 4.

Night falls and removes the possibility of further observation and photography. My return flight is set for tomorrow, so I head with my car for Memphis. In the dark I begin to superimpose the maps of Faulkner on those of Lins do Rego, to imagine them as neighbors, fellow *caribeños* with Glissant somewhere in between, describing the same general social history and geography, the same discourse of the plantation, from different viewpoints, and in different languages: I replay in my mind the visit to the Lins do Rego plantation in Pilar, Brazil.

New texts and the road to Pilar

A few days before my departure for Paraíba, the scene of Lins do Rego's sugar-cane cycle, a small, narrow strip of a state in the Northeast of Brazil (a region which also includes Bahia and Canudos), it shows up in the news: the governor of the state, Ronaldo Cunha Lima, had walked into a restaurant in the capital where his political rival, ex-governor Tarcísio Burity (whose name I would encounter on several publications about Lins do Rego, published with the financial help of the state government), was having a nice meal, and had blown him away with a '38 special. It seems that Burity and his political machine had been publishing more than literary criticism: the media under his control had denounced Cunha Lima's son for corruption in his administration. Cunha Lima would have ignored an attack on his own person, but the rules of medieval chivalry required him to "wash his son's honor" with the blood of his attacker. Here, then, is the escape from globalization I so fervently desire. Lins do Rego's region holds to its traditions, its honor and its shame. Paraíba is obviously the land of real politicians, who do the most necessary and important tasks themselves, and in the flesh, rather than depend on intermediaries and toadies or wait for that final settling of accounts in the gossip of eternity. Lins do Rego (1896-1957) was descended from a landed family of the area around Pilar that produced its own set of politicians. He grew up on a sugar plantation, the Brazilian term for which is *engenho*, studied law, and made his living from literature.

When I arrive at the capital of Paraíba state, João Pessoa, a few weeks later, Cunha Lima is still its governor. He has neither been indicted nor removed from office for his undeniable crime. Governors cannot be prosecuted. In Lins do Rego's works, João Pessoa is still called by its original name of "Paraíba." The name changed in 1930 because of a shooting entirely similar to this one. The governor, João Pessoa, defamed his political rival by publishing his correspondence with his lover in the local papers. The rival walked into a bakery and shot Pessoa dead. The crime was attributed to political motives, and helped unleash a revolution that shook Brazilian society to its foundations. One might find it odd that Lins do Rego never, in his work, mentioned the assassination of João Pessoa, the Constitutional Revolution of 1932, or the populist dictator, Getúlio Vargas. Censorship, real or anticipated, may be one reason, but another may be that politics for him was linked to family. Just as the natural landscape only appears in Lins do Rego's work as seen through the eyes of an actor in a human drama, so politics can only appear through the family. As Linda Lewin has pointed out, politics in Paraíba is a family affair: "Reliance on lineage claims for the exercise of political authority [is an] aspect of Paraíba's political system bequeathed by the conquest and settlement period. Such claims, which were usually inseparable from claims to landed property, asserted political prerogatives by virtue of membership in a 'first' or 'traditional' family. In the first place, this meant invoking racial criteria to demonstrate a politician's 'whiteness'—his real or putative Portuguese descent" (61). Thus, the most direct political depiction in the sugar-cane cycle, that of the strikes in Recife in *O Moleque Ricardo*, is presented in such a manner that no one without a history book—and I mean a local history book of Recife—would be able to understand it. The events are seen through the consciousness of the central character Ricardo, who has an incomplete grasp of the context. On the way up the coast towards its eastern end, I make stops in Salvador, where I learn more about Euclides da Cunha, and in Recife, capital of Pernambuco. Recife appears to me exactly as Lins do Rego describes it. Lins do Rego is concerned neither with old churches and convents, nor with the fortifications, nor with the pleasant beach at Boa Viagem, nor with the intellectual center in Casa Forte-Pipopucos, where his friend Gilberto Freyre set up his Institute Joaquim Nabuco for the study of Northeastern culture, and where I make use of the library. He is concerned with the business center of the city and with the "subúrbio." This cognate of English denotes the same peripheral phenomenon of housing located at some distance from a city center, but its connotations are exactly the opposite: poverty; discomfort; low housing values; and public transportation.

Recife, like most of the capitals of the Northeast, is a metropolis without industry. It merely receives all the relations and conditions of the interior of the state. The feudal lords take up residence in the capital: "The break-up of the rural patriarchy proceeds in rhythm with the development of the urban centers where, shortly after abolition, the free population migrated. The big house of the sugar plantation ['engenho'] is substituted by the two-story house ['sobrado'], where the seigneurial class coming to the city ensconces itself" (Landeira 481). Lins do Rego moves here from Paraíba to study and later to practice law, and, once his literary career is firmly in hand, eventually settles in the metropolis of Rio de Janeiro. My road to Pilar was Lins do Rego's road away from it. Some critics have called *O moleque Ricardo*, which describes the living conditions of the workers of Recife—most of whom, like Ricardo, have fled the oppression and poverty of the agricultural interior—the first proletarian novel in Brazil. Lins do Rego describes only the areas around the train tracks where the poor people live. I pass these in the first of the two legs of my bus ride to the library, and I see their overflow into the city center. His most evocative description is of the "mangue," that is, the arms of the ocean that reach inland, the water rising and falling with the tides, providing the small crabs that supplement the meager diet of the proletariat.

When I ask the taxi driver to take us to the bus station for our trip to João Pessoa, he stops every hundred yards and begs me to let him drive us there. Since I've had more expensive cab rides in Manhattan, I eventually give in. In our guided tour from Recife north to João Pessoa, we pass fields of sugar cane being cut by hand, and the small mud houses where the workers live. We pass through Goiana, near the border of Pernambuco with Paraíba, mentioned in *Menino de engenho*, the first book of the sugar-cane cycle, as a place of fevers and failed ambitions: "All the residents of the other house were prone with fever. They had returned from Goiana yellow and swollen with malaria.—Have the boy get quinine on the *engenho*. You leave here healthy and come back on death's door. Go ahead, give Goiana another try" (my translation; "Noutra casa o povo todo estava caído de sezão. Tinham voltado da várzea de Goiana amarelos e inchados de paludismo.—Mande o menino buscar quinino no engenho. Vocês saem daqui com saúde e voltam assim em petição de miséria. Vão outra vez para Goiana"; 28). For Ricardo, the disease, misery, and violence of the city. For those seeking work in the sugar mills, malaria. These are the invisible fences built around the boundaries of the feudal *engenho*. Those who climb them and escape southward resemble US blacks fleeing northwards—a phenomenon of geographical fate that also appears in Faulkner's work. There is indeed a sugar mill in Goiana, with a little chapel by its side. It had stopped producing sugar years ago.

The story Lins do Rego tells in his sugar-cane cycle is simple, straightforward, and to some extent autobiographical. The *engenhos*, whose owners were feudal lords and masters of all they surveyed, were gradually replaced by capitalist sugar mills, whose advanced technology allowed both faster and more refined sugar production. The autarkic and pluricultural sugar plantations, which began by selling cane to the mills, were eventually absorbed or, as recounted in the novel *Usina*, converted into mills. The Lins family was intimately involved in this transformation. The first sugar mill in Paraíba was constructed in 1917 on the former plantation Pau d'Árco, by Gentil Lins (Mariz 177). The capital investment necessary for the mill made it vulnerable to fluctuations in sugar prices. Between 1924 and 1931, world sugar stocks nearly doubled, from 16 to 28 weeks; this meant, of course, that sugar prices fell in nearly the same proportion. Some mills survived the shock; most *bangüês* (small *engenhos*, such as the Engenho Corredor where Lins grew up) did not. The crisis of the 1920s and 1930s confirmed the readjustment of the Brazilian economy, which had been in process since the foundation of the republic, from the sugar-producing north to the coffee-producing south. Cotton replaced sugar as Paraíba's main export. As a scion of (white) plantation aristocracy, Lins do Rego's path out of economic decline was more comfortable than that of dark-skinned Ricardo. He went to boarding school in Itabaiana, an experience he later shaped into the school novel, *Doidinho* (1933), studied law in Recife, and entered government service. He became instantly famous with his first novel, *Menino de engenho* (1932, my translation; "*Plantation Boy*"), which remains a Brazilian classic, as confirmed by several film adaptations and a comic book version. (It has also been translated into many languages.) He lived most of his remaining days in Rio de Janeiro, producing one novel per year, a rate he was able to maintain by eschewing all revisionary stages. A feeling of looseness and redundancy pervades most of Lins do Rego's novels. In *Fogo morto* (1943; "*Fire's Out*"), however, this redundancy is brought to perfection and becomes identical with the psychology of the characters. The redundant, meaningless chatter of people walking up and down the road past the house of the shoemaker, mixed with other sounds like the pounding of leather, the tingling of carriage bells, and the song of the blind beggar, absolutely captivated me when I opened the first pages of *Fogo morto* many years ago. I am on this road because I want to view the orchestra behind this symphony.

Late in his career, when he had run out of ideas for novels, and was enjoying the life of a literary patriarch and member of the Brazilian Academy, Lins do Rego retold in critical and historical essays the personal story he had spent six novels on. He himself had experienced the shifting of economic power away from agriculture towards industry, and he now made alarmist references to a "threatened culture" in his writing. All the leitmotifs of the sugar-cane cycle reappear in Lins do Rego's essay on the poet Augusto dos Anjos: the plantation in debt, the husband with no taste for the land, a mad uncle, above all the sensitive adolescent, the ex-servant living in the signeurial mansion, and final sublimation into the sugar mill. Lins do Rego strews stanzas of dos Anjos throughout his account. The latter's poetry is to be explained not through comparison with other *fin de siècle* literature, but through experience. The account is biographical, but at the same time moves beyond biography into geography. For Lins do Rego, the decadence of the sugar economy is not a matter of management or technology, but of blood and will. The sprouting of literary descendants is a final blow: "The Carvalho family had exhausted itself in the literary dos Anjos family" ("Augusto dos Anjos" 16). Lins do Rego chose to explain the poetic production of Augusto dos Anjos by evoking the place and plantation culture into which he was born: "[Anjos] was a melancholy boy of the Pau d'Arco plantation. The mansion was immense, with many rooms, slave-quarters at its side, waterwheel down below, the sugarcane fields in the alluvial plains, and on the hillsides the uncultivated land, where the red *paudarco* flowered in October and the yellow *paudarco* in November" (4). Memoirs that read like novels, novels that read like memoirs, true places and true imaginary places, the degree of fictionality remains relatively unimportant, as long as memory can still map the culture whose lifeblood is trickling away, whose major chroniclers can only survive in exile.

October and November are hot months below the equator, almost as hot as the February day when the taxi deposits us and our bags at the Hotel Central in João Pessoa. João Pessoa does not have the appearance or feel of a capital city; it is not rough and ready like Recife, but provincial. There are few tall buildings, the traffic is heavy only in two or three principle arteries; although the sun rises here earlier than anywhere else on the American *terra firma*, still by 8:00 a.m. commerce has barely opened and the streets are deserted. There is no touristic infrastructure, and everywhere I go people ask where I am from, as if they are expecting an answer that would allow them to mention relatives living in the same place. Tambaú, the beach eight kilometers from the center of town, is different, as the tourists congregate there to worship sun and sea. João Pessoa, the city itself, is left to the natives. João Pessoa is a lacuna in the work of Lins do Rego; doctors and other officials are always coming from the capital to treat matters on the Engenho Corredor—I estimate that the train ride took an hour, before replacement with the more inconvenient bus that takes two—but its physiognomy never appears in any of his books. The amount of begging, which has steadily increased as we travel northwards, reaches a crescendo in Paraíba. Here, begging is integrated fully into the economic structure. The beggar is the noise or parasite created by the system and helping to maintain it. The economy, still essentially feudal in conception, divides itself into fiefdoms. Capitalism is only a veneer over a much older way of organizing society. The open-air market predominates over the store system that surrounds it. The tents of the itinerant vendors are set up in front of the stores, making the town into an eternal bazaar (see Figure 14).

Even the government systems of assistance associated with capitalism have here instead the feeling of other fiefdoms, of personal touch. Elected representatives will be called on to furnish crutches and wheelchairs in return for the votes of the unfortunate. Votes are coerced or bought rather than won through propaganda. The government, as the largest employer in João Pessoa, has replaced the individual plantation owner as the voters' instructor. In *Fogo morto*, there is an enormous discussion about whom Vitorino will give his vote to. He insists on his independence in a system where alliance is everything, and pays for it with a beating at the jail (the jail still stands as a kind of museum in Pilar. It gives an excellent perspective over the rooftops of the town).

The beggar has his territory. He offers his wares and awaits a response, taking donation or negation with equal calm and indifference. The responsibility for his fiefdom keeps the beggar of the Northeast from becoming vicious. He does not smell of *cachaça*, as do beggars elsewhere in Brazil. His patron can be sure that his money is being well spent, vicariously. As I buy our bus tickets to Pilar, the beggar waits patiently for the transaction to be concluded before collecting the loose change. In his book of childhood memories, *Meus verdes anos* ("*My Salad Days*"), which is just

Figure 14. Mobile vendors ("camelôs") dominate the streets of João Pessoa, Paraíba

as concerned with depicting the true imaginary place of Pilar as are his novels, José Lins do Rego does not erase the beggar from his map of the plantation. The beggar is one of those unstable elements in the world, like the smugglers of *cachaça* and the traveling salesmen, condemned to wander on the road between Pilar and São Miguel, passing by the Engenho Corredor. Lins even notes the particular wares that the beggar has to sell: "The poor folk who begged for alms at the markets used to stop at the door of the mansion. The blind man Torquato with his guide. He used to walk the highway with a quince-tree cane, with the boy holding one of its ends. He recited his petition in a pungent tone. And he always brought Our Lord in to pay for others' charity. 'May God give you glory.' The glory of God! The glory of God must have been the kingdom of heaven, the life of a rich man in the world beyond. For a few pennies worth of alms, Torquato would offer limitless grandeur" ("os pobres que pediam esmolas nas feiras paravam na porta da casa-grande. O cego Torquato com o seu guia. Andava ele pela estrada com uma vara de marmeleiro, com o menino a segurá-la em uma das pontas. Cantava o seu peditório num tom pungente. E sempre trazia Nosso Senhor para pagar a caridade dos outros. 'Deus lhe dê a santa glória.' A santa glória! A santa glória devia ser o reino do céu, a vida de rico no outro mundo. Por um vintém de esmola Torquato oferecia uma grandeza sem conta"; 51-52). The same Torquato appears in *Fogo morto* as one of the most pitiful victims of the police of the region. He is thrown into jail along with José Amaro and Vitorino, and beaten. Blindness gives Torquato the license to beg. José Passarinho, on the other hand, is more like the proletarian beggars of the South: he requests cigarettes and booze. In return, however, he becomes José Amaro's housekeeper when his wife leaves him. Like Torquato, Passarinho rewards donations with exaggerated language. José

Amaro is a beggar disguised as an "oficial" (that is, someone with an office or trade; Amaro is a saddler). The crisis of the novel is precipitated when Seu Lula, owner of the land on which Amaro's house is situated, banishes his tenant. José Amaro must beg for his little piece of land in this latifundarian world in which he thought he had a place, and the prospect of wandering the roads of the area like Torquato and the other beggars terrifies Amaro into committing suicide. Of the major figures in *Fogo morto*, only Seu Lula and Vitorino Carneiro da Cunha never beg. The two of them, at the extremes of poverty, never have to lower themselves to that device. The weight they place to counter their poverty is the same: family. Family, the clan, while important throughout Brazil, is perhaps most important in the Northeast, where it controls everything and the individual nearly disappears under its weight. Whenever the young Lins do Rego of *Menino de engenho* does anything abnormal, the household immediately attributes the *faux pas* to his father's blood, not to his own inexperience or misguided will. The governor tried to kill his rival not to save his own honor, but his son's. The analysis of *Meus verdes anos* in terms of family relations would require a book in and of itself, with diagrams of kinship relations à la Lévi-Strauss, so complex are they. The clans are not one big gooey mess of warmth and mutual assistance, but balancing acts between love and jealousy, economic and political one-upmanship.

The city bus takes us first to the resort community of Tambaú, where I deposit my family for sun and fun. I then get another bus back to the José Lins do Rego Cultural Center. The structure is a huge roof without walls, allowing the breeze to blow through. Underground is a library and the museum honoring José Lins do Rego. There I speak with the director, Dona Ylusca. She turns out to be the cousin of Rómulo, a Brazilian doctor of our acquaintance living in Altoona, Pennsylvania. Escaping northward is also an option, then. When I tell Ylusca that I would like to visit Corredor plantation, she discourages me from doing so. There is a problem of "inventário" (i.e., a dispute over inheritance). That is the story, at any rate. The museum's center is José Lins do Rego's personal library. His books, his shelves, his chairs, and his tables, a portrait of his friend, Gilberto Freyre, all are reverently placed here (see Figure 15). My eyes scan the shelves for any Faulkner that might have been part of Lins do Rego's library. I find only a Portuguese translation of *Requiem for a Nun*, too late to have influenced *Fogo morto*. Next stop, Pilar.

From João Pessoa it is an hour by bus to Pilar. We pass by extensive open fields, first of sugar cane, and then of pineapples. Not a sign of the cotton that became Paraíba's economic staple, after sugarcane had declined. In Lins do Rego's works, cotton is grown by the inhabitants of the plantation and sold to the owner. Anyone caught selling their cotton in Pilar or to another plantation is first threatened with having their terrain invaded by the bovines of the plantation, who will exterminate their crops. Then he is threatened with exile, easily accomplished if things reach that point. Exile is the fate suffered by José Amaro.

The driver knows where Engenho Corredor is located, and the bus drops us at the scene of *Menino de engenho* and *Meus verdes anos*. We can see the houses on the outskirts of Pilar, looking small at a distance of two or so kilometers. The mansion of the plantation is hidden from our view, behind mango trees in which a group of small monkeys frolics. A dog appears on the porch of the mansion; everything is silent, as

Figure 15. Dona Ylusca in José Lins do Rego's personal library and study.

strips of white cloud pass over a continually changing sky. My wife claps at the gate, but the house is far away, and no one hears.

My daughter begins to cry, asking why we came to this place with nothing to do and no one in sight. We finally ask a passing bicyclist if this is Corredor. He says that it is, and points the way in. We reach the big house, and find a receptive welcome from the lone housekeeper. Unrelated to the Lins family, she simply guards the mansion and keeps it from further decay (see Figures 16 and 17). Evidence of the long-departed sugar industry is everywhere: large metal vats; a huge, empty courtyard where the sugar used to lie before being ground for its juice; and huge metal spoons and canisters everywhere. Everything has survived where it was thrown down, as if awaiting elevation to the protected status of Faulkner's Rowan Oak. A row of small rooms set off from the main building looks suspiciously like the "rua" (literally, street, the slave and later servants' quarters), but I am assured that it is not. The housekeeper is happy to show us all

Figure 16. Big House of the Engenho Corredor, Pilar, Paraíba

Figure 17. Big House seen from the inner courtyard

the rooms, mentioning the one they *think* Lins do Rego was born in. She mentions that an old lady who was Lins do Rego's nursemaid lives on the other side of the river.

Our long walk into Pilar follows the route—or so I imagine—of the parade of people of all classes past the door of José Amaro. Soon the dirt road turns into cobblestone, and long rows of houses appear on either side. The same stillness and absence of people prevails. We visit the jail, scene of the beatings in *Fogo morto*. Then we cross the river and start back up the other side, in search of the old lady, last eyewitness of the past. I'm not even sure why I'm looking for her, or what I will ask her if I find her. We pass the deserted train station that used to carry people and goods back and forth to the capital and go on, crossing several farms. We finally realize that this is taking too long, and turn back. On the way, a young man on a bicycle greets us. José Aurélio de Melo, a university student, is preparing a video biography of Lins do Rego. He denies the existence of the old woman. Doing some simple arithmetic for the first time in my excitement, I realize that Lins do Rego's nursemaid could not possibly still be alive. On the way back, he shows us the house where Lins do Rego grew up, and where he, José Aurélio, is now living.

The rusting machinery, the tall tales, the chance encounter, the jail where the men are beaten. I have had my backstage tour of the world of the novels. As we return from Pilar in the five o'clock bus, swapping mental maps of the area with our new-found colleague, José Aurélio, who needs to attend class at the University of Paraíba in João Pessoa, night begins to fall. Night comes here early and there is little twilight. "Never were there so many daybreaks and nightfalls in a Brazilian novel [as in Lins do Rego's works]," says Virgínius da Gama e Melo, with some justification (127). The early mornings leave their impression; the early nightfall is beneficial, given the discomfort experienced in the full sun. In the night all cats are gray; it gives relief from the constant maintenance of social roles of the day.

Tryptichs of solipsism, *Fogo morto*, and *The Sound and the Fury*

Imagine a map in which each border, boundary, and topographic feature spoke of its delimitation through human vicissitudes. These are the mental maps of Faulkner and Lins do Rego. Both authors devoted their work to the issue of how present geography and social conditions flow from the initial conditions of colonization, slavery, and a patriarchal seigneurial system. Both authors took seriously the task of using a relatively small "postage-stamp of soil," in Faulkner's words, as a metaphor for nations that are composed of differences—between North and South, rural and urban, and black and white. A comparison of the two authors reveals similarities that make them belong to the same region, a true imaginary Caribbean, across national and linguistic boundaries; and, as Gabriel García Márquez suggests, their mental maps extend the region far outside the borders placed on the region by conventional geography: "[Faulkner] has seemed to me to be a writer from the Caribbean [. . . which] is not a geographical area around the sea, but, rather, a more vast and complex region, with a homogeneous cultural composition, which extends from northern Brazil to the U.S. South. Including, of course, Yoknapatawpha County" (García Márquez n.p.). "Region," then, plays out its double meaning in the work of these authors, as both intra-national and international. Faulkner seemed to pay tribute to these similarities by having Thomas Sutpen, the protagonist of *Absalom, Absalom!* (1936) begin his career as a planter in the West Indies.

John T. Matthews notes, however, that the Caribbean origins of Sutpen's project began to be considered in readings of *Absalom, Absalom!* only in the 1990s. Before then, Matthews contends, the non-US origins of the plantation system were a critical blind spot (or, in his terms, a fetish): "Like its narrators, readers of *Absalom* . . . have always had before their eyes Faulkner's evidence that the plantation South derives its design from new-world models, owes a founding debt to West Indian slave-based agriculture, extracted labor and profit from African-Caribbean slave trade, and practiced forms of racial and sexual control common to other hemispheric colonial regimes. But there is a kind of knowledge that can be held while being ignored, a kind of vision that looks but does not see" (239). As Antonio Benítez-Rojo has suggested in his book *The Repeating Island* ("*La Isla que se repite*"), an understanding of Caribbean culture should be based upon a "discourse of the Plantation." The ubiquity and shaping influence of this "discourse of the Plantation" is demonstrated in the comparability of these two novels, even if only one of them actually takes place on a plantation. Faulkner's Compsons do not engage in agricultural activities; yet the values, social hierarchy, in short, the mental maps by which the characters navigate their world derive from the discourse of the plantation no less than they do in Lins do Rego.

Da Cunha and Benet, natives of the metropole, give a sense of landscape to their nations' regions viewed from the outside, as it were. Landscape provides the visible clues for a "terra incognita" or "black box" of social and psychic systems that lie outside the ability of the observer to familiarize. The "insiders," José Lins do

Rego and William Faulkner, scions of slavocratic families both, on the other hand, provide a sense of "place." What José Castello says of Lins do Rego could apply to Faulkner as well: "Regionalism for him is not simply photography of features or characteristics of a region. It is much more. It is the heartfelt, profoundly human and lyric testimony of nature and of human conditions under telluric contingencies brought about by economic and social transformation" (189). The mental maps examined in this chapter are drawn along the lines of human relationships, especially as these emerge out of the conflict with racial, class, and gender distinctions. Such relationships, as my chapter title implies, divide as much as they unite.

Although Brazil and the United States liberated themselves from European political domination relatively early, neither nation upon independence rejected either the inherited colonial system of slavery, deemed the most efficient for producing staple crops (sugar and cotton, respectively) for the metropole and an international market, nor the pseudo-feudal system of social relations engendered by the plantation mode of production. The cataclysmic violence of the American Civil War, versus the relatively peaceful abolition of slavery in 1888 in Brazil, may hide more fundamental similarities: the American South and the Brazilian Northeast had by the end of slavery already seen their economic dominance superseded by other regions, the industrial Northeast in the case of the United States and the coffee-growing states of Minas Gerais, São Paulo, and Rio de Janeiro in the case of Brazil. In both regions, abolition did not fundamentally change the relations between masters and former slaves. Rather, "whereas the *senhor* previously had been able to extract the surplus product from the laborer by virtue of his ownership of the laborer, he now did essentially the same thing by virtue of his ownership of the land." (Taylor 64). Numerous scholars, among them Deborah Cohn, Harley Oberhelman, Daniel Richardson, and Lois Parkinson Zamora, have sought to overcome the divides of linguistic and political boundaries to reconstruct the aspects shared between American literatures. Fred Ellison has adumbrated the similarities between the Deep South of the U.S. and the *Nordeste* of Brazil: "Like the [North American] South, the [Brazilian] Northeast was during the colonial era the center of a rich agricultural economy based not on cotton so much as on sugar cane, and its society was an aristocracy of large landholders marshaling armies of slaves. On these foundations arose a peculiarly stable culture which produced some of Brazil's greatest statesmen, scholars, artists, poets, and novelists" (3). Eugene Genovese agrees that slaveholders in the U.S. South and Brazil closely resembled each other, and that both groups approximate most closely the "standards of paternalism we associate with the patriarchal plantation" (96). The agricultural basis of the Southern economy, as well as the inhibitions to industrial development caused by the Civil War and by Reconstruction, are too well-known to bear repeating here. Published one year after the appearance of *The Sound and the Fury*, the famous collection of Southern Agrarian essays, *I'll Take My Stand*, focused on the difference between the "American," commercial-capitalist economy and the South. The collection's opening "Statement of Principles" notes that all the essays "tend to support a Southern way of life against what might be called the American or prevailing way; and all as much as agree that the best terms in which to represent the distinction are contained in the phrase, Agrarian *versus* Industrial" (Rubin xxx-vii). The analogous oppositions, South/American and Agrarian/Industrial, recast the familiar conflict between South and North, Confederate and Yankee, in terms that make the North both a region and a nation.

Similarly, but in the opposite direction, inhabitants of the Brazilian Northeast could observe the center of economic power descending southwards, particularly at the end of the nineteenth century: "By 1872, in terms of total regional income, the entire northeast had about a quarter of that of the provinces of Rio de Janeiro and São Paulo. By 1900, it had perhaps a sixth of their total, with the impressive growth of São Paulo as a critical aspect of the northeast's slide" (Greenfield 40). *Nordestinos* saw this shift as a defeat of their culture and way of life. It is no accident, as Giorgio Marotti points out, that "the years of [Lins do] Rego's literary career coincide precisely with the [Getúlio] Vargas era. . . . Under Vargas occurred the beginnings of transformation and industrialization against the world of preservation that for centuries had weighed down the agricultural fiefdoms. . . . In short, another Brazil was emerging into the foreground and historically overtaking and partially excluding the world to which José Lins do Rego belonged" (268).

Lins do Rego told the story of the loss of power and prestige by the ruling elites of the Northeast most powerfully in the novel *Bangüê* (1934), the third work of the so-called "Sugar-Cane Cycle." Carlos de Melo, the protagonist of the first two novels, inherits the family plantation and proceeds to lose it to José Marreira, a descendant of slaves. Marreira triumphs through a combination of hard work, strategic political and economic alliance, cunning, and the camouflaging of his intentions beneath a mask of Uncle Tom-like submissiveness to Carlos's superior social position. As either allegory or example of the decline of the Northeast, the novel is historically misleading: few if any plantations in the region were taken over by blacks, who remained politically, socially, and economically disenfranchised. The racial paranoia of *Bangüê* has led some critics to describe the author's ideology as reactionary and even "backward" (Marotti 324). However, there remains some ambiguity due to the fact that the novel never pretends to describe reality, but only the mental map its protagonist Carlos draws of the racial and economic landscape. The novel is in this sense propadeutic to the three fractured visions of the more accomplished *Fogo morto* — as well as to Quentin Compson's obsessions. Lins followed the theoretical pathway laid out by Gilberto Freyre in his *Manifesto* of 1926: Northeastern *Regionalismo tradicionalista* ("traditionalist regionalism") opposed itself to the European-influenced *modernismo* centered in São Paulo and other urban centers of the Brazilian South. In particular, Lins wished to use a language more attuned to the rhythms of daily speech, a "geopolitical language" different both from the more elevated, generically literary style of previous regionalists (such as Euclides da Cunha), and from the "researched and invented" language of the modernists (Ivo iv). *Fogo morto* is in one sense the culmination of Lins's ambitions in this direction, an almost complete surrender of narrative control to the "native speakers" of his drama. Normally used to signal the end of the cane harvesting and milling period (the "fire" of the title being that used to separate and process the cane syrup). The "morto" of *Fogo morto* literally means "dead." Lins's title refers as well to the passing of the sugar-cane plantation as a way of life. This social death impresses itself upon everything in Lins do Rego's work. The first element of the author's mental map, reminiscent of Gogol's, is the regulation of everything by the principle of life and death, which is the only triumphant element of his works. All of Lins do Rego's memorialism records only this single fact of the vitalization of death: "Even the physical changes wreaked on the landscape through the social find . . . a guarantee of life, up to the time when a totally dead and mummified world, as happens when the mill substitutes the plantation, the socio-economic repercussions of which will transform the multicolored landscape into one of canefields of pure green" (Gama e Melo 75). Even the seeming destruction of nature through the sugar mill becomes part of the natural vegetative process.

It is indeed a colorful world; colors designate social classes. Social relations are a complex grid in which are mixed categories of race, economic status, family relations and descent, and social power. As in Faulkner, Lins do Rego shows racial lines and barriers as one feature of a complex landscape. The categorizations of people in Lins do Rego's work are indefinite and constantly shifting, despite the seeming rigidity with which they are upheld. Even an "easy" term like "branco" ("white") does not indicate a skin color. In speaking of the bandit Antônio Silvino, Lins gives the real definition: "The black women thought the bandit a gentleman, a white man" ("As negras acharam o bandido um homem de tratamento, homem branco"; *Meus verdes anos* 75). To be white is to be someone who must be reckoned with. So too Vitorino's wife, who equates the term with goodness: "A trickle of blood ran down his face. He was a white man, a good man, an unreasoning child" ("Corria um fio de sangue de seu rosto. Era um homem branco, era um homem bom, uma criança sem juízo"; *Fogo morto* 200). Vitorino uses the term in a similar way in order to prop himself up in his conversation with Lula: "Vitorino got up and showed no fear. 'No one yells at me. I'm as white as you are, Colonel'" ("Vitorino levantou-se, e não se amedrontou.—Comigo ninguém grita. Sou tão branco quanto você, Seu Coronel"; 200). And the opposite term, "black," does not designate skin color as much as position of servitude. Vitorino again: "I'm not your negro. Go talk that way with your field-hands" ("Não sou seu negro. Fale assim com os cabras de sua bagaceira"; 172). Again, his wife makes a similar equivalence: "'Mrs. Adriana, any news of your son?' She had to speak, to ingratiate the black woman. . . . And she departed furious with the black" ("Tem tido notícia de seu filho, Sinhá Adriana? Teve que falar, que fingir satisfação para a negra. . . . E saiu irritada com a cabra"; 34-35). So a *negro* is also a *cabra*, but what is a *cabra*? Here the fun begins. A standard dictionary gives us four different although related meanings: one is a racial mixture. Another is a *capanga*, that is, the hired hand of a *senhor*. Another is a *cangaceiro*. This third meaning is close to the previous one, since a *cangaceiro* is a senhor without an *engenho*. And in fact, the followers of Antônio Silvino, a bandit who appears as a symbol of freedom and terror in the novel, are constantly referred to as his *cabras*. Another meaning is that of any rural inhabitant, or, more properly, anyone who lives in the fields. We can detect in these four meanings three parameters of categorization: racial, social, and economic. It seems to me from the examples given that the word may be used in all three of these registers at once.

Both authors are simultaneously critics and nostalgists of their respective regions. Faulkner, for example (himself a rider), tends to identify the horse with fading aristocracy (see Milum), and the automobile with progress and a loss of values, while the main theme in Lins do Rego's early work is the struggle to preserve the traditional *engenho*, or sugar plantation, against the *usina*, or sugar mill. Mariana Soares has compared Lins do Rego's works with those of Albert Camus and other existentialist writers. Both authors present their respective regions as lacking in feeling and human communication. The social results of the decay of plantation society in the *Nordeste* and in the Deep South overlap with the urban alienation and *anomie* that gave literary modernism many of its themes, thus reproducing on the psychological level the game of identification-disjunction of each chosen region with the nation as a "whole." My comparison of these two authors—or, more precisely, of their masterpieces, *Fogo morto* and *The Sound and the Fury*—treats a less expected similarity in the style and structure of their works, through which we may glimpse a different kind of regionalism, the effect of history on consciousness. For reasons to be investigated in the following, both authors choose to refract their regions through the consciousness of several characters rather than just one. In both works, "regionalism" operates as a linguistic process, an exploration of the construction of the mental maps of racial, social, and gender division through words.

Perspectivism results in a triptych of interior monologues that comprises each work. *Fogo morto* (hereafter *FM*) is divided into three chapters: "O mestre José Amaro" ("The Leatherworker José Amaro"); "O Engenho de Seu Lula" ("Mr. Lula's Plantation"); and "O Capitão Vitorino" ("Captain Vitorino"). *FM* stands out from Lins do Rego's other works in the musicality of its repetition of themes in somewhat rigorous form. The poet, critic, and musicologist Mário de Andrade referred to the internal coherence of this structure of separate monologues with the metaphor of sonata form: "*Fogo morto* achieves precisely the form and spirit of the sonata, as it deals with a single psychological datum, a single 'business,' the superiority mania, treated in three themes, three melodies, three movements" (292). This superiority mania can be seen in the fact that each chapter title involves a personal name linked to its social designation. Lins's psychologism tended towards repetitions of his protagonists' obsessions; his previous novels of the sugar-cane cycle (*Menino de engenho*, *Doidinho*, *O moleque Ricardo*, *Bangüê*, and *Usina*) had generally taken the form of memorialism, and read as if the author were directly communicating his thoughts and memories onto paper. Gilberto Freyre had suggested exactly the refinement of technique encountered in *FM* in a letter to Lins dated 29 November 1938. Freyre wrote, "If I were you I would not let myself get carried away with this idea of being prolific and would concentrate instead on putting more time into producing one novel without the mechanical repetition of effects, motifs, etc. A novel with a deeper psychology" (Freyre n.p.).

Although Faulkner's possible influence on Lins do Rego seems to me only a partial answer to the reasons for the similarities in their works, I have devoted some time to investigating the question. As noted in the preceding interlude Lins do Regos's personal library, which has been transferred complete to the Museu José Lins do Rego in João Pessoa, Paraíba, contains only one Faulkner novel, a Portuguese translation of *Requiem for a Nun* (*O Mundo não perdoa). The Sound and the Fury* did not appear in Portuguese until 1960, too late to help Lins, whose English was not good enough to read American literature in the original. There remains, however, the chance that Lins do Rego read someone else's copy of the celebrated French translation by Maurice Coindreau, or that he learned of the work through conversation with someone who had read it.

The Sound and the Fury (hereafter *SAF*) is divided into four chapters, each headed by a date. Three of these chapters are internal monologues (by the three Compson brothers, Benjy, Quentin, and Jason). These correspond to *FM*'s chapters in ways I will make clear in what follows. Faulkner's fourth chapter, centering on the Compson's servant Dilsey, is narrated in the third person. Faulkner's triad has often been compared with the Freudian id, ego, and super-ego, Lins's less so. In such schemes, Benjy represents of course the id, Quentin the ego, and Jason the super-ego or repressive mechanism. It is readily apparent that the solipsism of each character provides a fragment of a "normal," integrated personality. Were they to come together into a single personality, the three male characters of each work could balance each other's eccentricities. In their divided, solipsistic existence, each character represents a particular incompleteness that turns into madness. In addition, each character represents a particular social position, and when taken together provide a map of their respective places as "knowable communities." That is, the same triangulation of characters that implies the impossibility of a stable, coherent self simultaneously marks out the limited social positions available to individuals in the regions depicted.

Benjy and Seu Lula

Benjy is the youngest Compson, brother to Caddy, Quentin, and Jason. Born mentally deficient, the mute Benjy needs constant monitoring, undertaken mostly by the Compson servants. Lula Chacon de Holanda, owner of the Santa Fé sugar plantation, would seem to be Benjy's opposite. Yet both characters are helpless, their communicative abilities crippled by highly visible psychological defects. Lula, no less than Benjy, is an idiot in François Pitavy's reading of the concept: "The idiot . . . is the one without the Other, confined to the order of the imaginary, without the ability to rise to the order of the symbolic, since the 'I' is immutably related to an ideal self. Thus the idiot can be viewed as embodying idealism to an absolute degree, which can also be construed as the reverse of idealism . . . buried as it is in the limbo of consciousness" (102). Pitavy uses here the distinction made by Jacques Lacan, who based the concept of the imaginary on his finding that "the ego of the human infant . . . is constituted on the basis of the image of the counterpart (specular ego)" (210). The symbolic, on the other hand, is the pre-established order of language, based upon difference rather than similarity, to which the ego must commit itself in order to achieve normal development. As Pitavy's Lacanian vocabulary indicates, both Benjy and Lula represent the psychosis induced by entrapment in the imaginary and an inability to grasp the symbolic dimension of human interaction. Both characters are shown as imprisoned in a world of images. Benjy's, as Pitavy implies, are a stream of images linked to an emotional affect, while Lula's are more specific images of his three fathers—biological, in-law, and God the Father—against all of whom he measures himself and comes up short.

Son of a hero of the resistance to federal power, Lula personally represents the decline of the plantation owner to a parasite, living off the reputation of his father and the accumulated capital of his father-in-law. When he appears on the Santa Fé, the owner Captain Thomas is thrilled not so much by Lula's presence, as by the chain of images he brings with him. Lula is judged through his father: "[Lula's] father had died in the rebellion of 1848. Much was told of Antônio Chacon's courage, of how he had fought a government force in the jungle of Jacuipé until he died" ("O pai [do Lula] morrera nas lutas de 48. Contava-se muito do coragem de Antônio Chacon, cercado nas matas do Jacuípe, com Pedro Ivo, batendo-se com uma força do govêrno até morrer"; *FM* 127). Regionalism in this case opposes itself to the nation, produces a direct splitting off and opposition. The solipsism of regional resistance is transferred to the son in its psychological format. On the social level, then, both Benjy and Lula embody the inability of the inheriting generation either to carry on the traditions of the old, or to strike out in a new direction. Faulkner and Lins use the same symbol for this parasitism that eats away at patrimony: the trading of gold coins until they disappear. Faulkner represents this final bankruptcy in the novella, "The Bear." When Ike McCaslin is to receive his inheritance, he finds the urn filled with I.O.U. slips. Like Ike's guardian, Carothers, Lula also takes gold coins until none are left. When the bandit Antônio Silvino invades Lula's house in search of this hidden treasure, he refuses to believe Lula's assurances that all his gold has been traded away, and his threats trigger Lula's final mental collapse and his exiling of José Amaro, whom he rightly suspects of collaborating with the bandit. Faulkner chose to end his novel with a powerful image of stagnation: Benjy insists on always being driven around the statue of the confederate soldier to the right. When the inexperienced Luster takes him to the left, Benjy starts howling so loudly that his brother Jason is forced to come to the rescue. As the carriage corrects its course to the right, "[Ben's] eyes were empty and blue and serene again as cornice and façade flowed smoothly once more from left to right, post and tree, window and doorway and signboard each in its ordered place" (371). The images Lula lives by are religious icons before which he kneels, and which help him identify with yet another father figure, Jesus Christ: "When the canon Frederico would raise the golden chalice to the Lord, and the bells resounded in the church, Lula would feel victimized by men. The image of his father would appear to him . . . his past as the sacrificed one, son of a poor widow" ("Quando o Cônego Frederico elevava ao Senhor o cálice de ouro, e as campainhas ressoavam na igreja, ele sentia-se uma vítima dos homens. Aparecia-lhe então a imagem de seu pai . . . o passado de sacrificado, filho de viúva pobre. Sim, ele, Luís César de Holanda Chacon, não era o que deveria ser, fora roubado do que era seu"; *FM* 163). Lula's self-recognition—the recognition of a lack or abyss in the very center of his being—is stimulated by images and sounds. Christ's passion and death blend themselves with the passion and death of his father, and then with his own.

Benjy and Lula live in autistic worlds characterized by imagistic identification and self-referential language. Benjy's language is dominated by what Roman Jakobson calls the "emotive" function. "Caddy smelled like trees," the most persistent of Benjy's formulations, is his translation of the excessively abstract "I love Caddy," an emotive utterance par excellence. Here are a few examples: "Caddy smelled like trees" (*SAF* 48); "Caddy smelled like trees. . . . She smelled like trees" (49); "Caddy smelled like trees" (51); "Caddy took the kitchen soap and washed her mouth at the sink, hard. Caddy smelled like trees" (55); "I went and Father lifted me into the chair too, and Caddy held me. She smelled like trees" (82). Benjy's language deficiency is complemented by an acute sense of smell that serves as his tool of navigation and mapping more than does vision: he can, for example, smell his grandmother's death, and he wails at the smell of Caddy's perfume that obliterates her vegetable odor and indicates her passage into womanhood: "Caddy put her arms around me, and her shining veil, and I couldn't smell trees anymore and I began to cry" (46). The invocations of Caddy's smell become a leitmotif in the first chapter: These repetitions serve to atemporalize Benjy's world. Faulkner chose the past tense to represent the languageless atemporality of Benjy's mental world, using italics to differentiate between memory and perception. Thus, Benjy's inability to function in the present appears textually as a fixation on past events. Like Benjy, Lula is characterized precisely by his silences, by his inability to interact with the world. Lins do Rego chose to reproduce Lula's thoughts in an interior monologue only in certain strategic parts of his chapter. Mercedes Risso notes that Lula's interior monologues, unlike Amaro's, are characterized by what Bakhtin calls "divergent bivocal speech"; that is, his monologues always present both the word of the character and, at the same time, an ironizing oppositional word (31). Thus, the tension of this chapter is created by the gap between Lula's behavior and other people's explanations for it. In his earlier work, Lins do Rego had used the interior monologue almost exclusively. The narrator of his first three works, *Menino de engenho*, *Doidinho*, and *Bangüê*, gives us a very good idea of what Lula's thoughts might look like. Carlos, the narrator of these earlier works, is like Lula obsessed with the ideal image of the plantation owner and with his own failure to live up to that image. We can imagine sentences like the following coming from Lula's mouth, as he observes the portrait of his grandfather in the big house of the Santa Rosa plantation (which also appears in *FM*): "I went out into the living room and there was my grandfather's portrait hanging on the wall. My grandfather's good face, the sweet eyes, old Zé Paulino himself came alive within the frame. And if he were alive and well, the Santa Rosa plantation wouldn't be given up to anyone. It would be his. . . . I couldn't measure up to his club. His blood was not in mine. I was of another race, I was another's grandson" ("Sai para a sala de visitas e lá estava o retrato do meu avô pendendo da parede. A cara boa do meu avô, os olhos mansos, todo o velho Zé Paulino ficava vivo na moldura. E se fosse vivo e forte, o Santa-Rosa não seria entregue a ninguem. Seria dele. . . . Eu não podia com o seu cacete. O seu sangue não estava no meu. Eu era de outra raça, era neto de outro"; 190-91).

The works of Lins do Rego reveal no unifying gaze of the plantation owner surveying his property, as the text of Goethe provided. This scene shows us the transformation of such surveying in the decadent plantation: the gaze of a dead ancestor turned not outward onto the landscape, but back onto his own progeny, whereby he freezes them into inaction. Trapped in the imaginary, Carlos measures himself against the mirror image of his grandfather and comes up short. The Oedipal conflict is defused through the denial of blood relations. Interestingly, the figure of Lula also appears in these early novels, but featured only in his extreme poverty and his ability, like Zé Paulino, to hold on to his plantation and maintain his patriarchal presence to the very end. In *FM*, we learn Lula Chacon's history for the first time. Like Benjy, the linguistically crippled Lula finds his only stability in a world of sounds and images: the sad waltz his wife plays on the piano; the golden curls of his daughter Nenem; and the religious idols of his sanctuary provide his only refuges. Both characters anchor their fragile selves in imagistic identification rather than in linguistic communication. They symbolize that aspect of place that seeks to preserve itself unchanged by producing an impossible past and holding it up as a model.

Capitão Vitorino Carneiro da Cunha and Jason Compson

If Benjy and Lula represent psyches entrapped in the imaginary, Captain Vitorino and Jason Compson are fully integrated into the Law of the Father, or the symbolic order. Once again, the similarities between these two characters are hidden behind their exterior differences. Vitorino has no visible means of support, whereas Jason Compson is the only member of his household with a steady job, as he likes to remind himself and others constantly. Vitorino's major action in the novel involves his heroic efforts to free his "compadre" José Amaro from jail, and to reinstate him in the good graces of his landlord, Seu Lula. Jason's major actions are extorting money from his sister Caddy, and unsuccessfully trying to prevent Caddy's daughter Quentin from running off with a rounder (whereby the lovers also steal back the money Jason had extorted). The crucial similarity between these two characters is that both interact with their social surroundings in a way that none of the other characters examined are able to—and both accomplish this feat by repressing any interior voice or self-reflexivity. Like Lula's, their language echoes, but the echo of the Other's discourse which speaks them disguises the original utterance. Theirs is a word that derives from the Other, that is directed back at it, and whose purpose is to sanction what they do. It is only by exteriorizing their "selves" that they can maintain themselves. Faulkner is not being ironic in the 1946 appendix to the novel when he calls Jason the only "sane" Compson. "Sane" is determined by social norms, and Jason is the only Compson able to survive in society. Jason's "normalcy," compared with Quentin's suicide, Benjy's idiocy, and Caddy's homelessness, make him a precursor of the Snopeses, who in turn represent an invasion of the piney woods into the center of Yoknapatawpha.

There are moments when the masking with the Other's discourse fails, however, revealing the abyss separating each character's fragile ego from society's actual assessment of them. For example, in a society where first-name address is the rule, Vitorino always insists that others call him by his full name and title of "captain." However, throughout the text his name is echoed back at him by mischievous children as the ridiculous "Vitorino Papa-Rabo." If Lula's language is phatic, Carneiro's is "emotive" in that every word he speaks is intended to point to himself as speaker. Accordingly, the leitmotifs in Carneiro's chapter are the following: "Vitorino Carneiro da Cunha fears no one" ("Vitorino Carneiro da Cunha não tem mêdo de ninguém"); and "I have a name" ("Tenho nome"). All of these phrases are bounced off the addressee and returned to the speaker. The text contains one example after another of utterly useless dialogs between Carneiro and others. As with the other characters, communication is shown to be impossible; the difference is that Carneiro seems to consciously will these misunderstandings. Summoned by the lieutenant who will later imprison his friends, Carneiro manages to steer the dialog in a direction in which no information is exchanged: "'I'm not a man to be trifled with, lieutenant. . . . I only know one thing, lieutenant, there isn't a man alive who can make me run and hide under the bed. If he treats me with respect, then I respond politely; if he yells at me, I yell back. . . . You are an officer, and so am I. I have my commission, and I will die with it on the field of honor'" ("Eu não sou homem de deboche, tenente. . . . Eu só sei é de uma coisa, tenente, não há homem que me faça correr para debaixo da cama. Se vem com jeito, eu respondo com delicadeza, se vem com grito, grito com ele. . . . O senhor é autoridade, eu também sou. Tenho patente, e morro com ela no campo da honra"; 236-37). João Pacheco says of a similar passage that "the description is completely external. It is so lively, however, that in its texture one discerns Vitorino's internal world, invisible but present and tangible, in a state of ferment" (60). I would add that the external nature of description in this chapter is a signifier of the fact that Vitorino's interior is identical with his exterior. Unlike Amaro, for whom the pejorative term "lobishomem" ("werewolf") becomes a divided word, coming from outside but entering into his own interior monologues, Carneiro puts an absolute distance between his own full name and the phrase "Papa Rabo" that people constantly confront him with. This exteriorizing of the self is also present in Jason's section of *SAF*. Of the three monologues, Jason's is the one that most resembles spoken language. We can imagine ourselves in the drugstore, getting a "dope" (i.e., Coca Cola), prisoners of this man's unending complaints. The first three sentences Jason speaks in his section of *SAF* all include the verb "I say": "Once a bitch always a bitch, what I say. I says you're lucky if her playing out of school is all that worries you. I says she ought to be down there in that kitchen right now, instead of up here in her room" (206). And yet the role of the speaker in Jason's discourse is not personal, but collective. The first of these sentences is a *cliché*; Jason respects what the community says. His language, his thought is a representation, a condensation we might say, of the Law of the Father: correspondingly, his first sentence denigrates women. When Jason receives Caddy's letter a few pages later, the Law of the Father repeats itself: "I opened her letter first and took the check out. Just like a woman. Six days late" (218). In turn, this repression receives its just reproof at the end of the novel, when Jason's niece Quentin defeats her uncle. But we have hints throughout the text that the Law of the Father is not invulnerable. As he watches the dirt being thrown onto his father's coffin, Jason begins to feel funny: "watching them throwing dirt into it, slapping it on anyway like they were making mortar or something or building a fence, and I began to feel sort of funny and so I decided to walk around a while" (232). This funny feeling is the "mourning work" that Jason will not allow to be completed. A few minutes later, Jason encounters his sister Caddy, who has been exiled from home, standing at the grave. The feeling is repeated: "I didn't say anything. We stood there, looking at the grave, and then I got to thinking about when we were little and one thing and another and I got to feeling funny again, kind of mad or something" (233). Jason's anger, an engine driving his discourse throughout his chapter, is a secondary emotion driven by the repressed primary ones of love or sorrow that he cannot say, because they are not in the standard vocabulary of his community.

About Mestre José Amaro and Quentin Compson

As with Benjy and Lula, it would be hard to imagine two characters in more different social positions than José Amaro and Quentin Compson. Amaro is an uneducated saddlemaker of uncertain origins. He passes his entire life in a house on Lula's plantation. His position at the edge of the road to Pilar represents his liminal social status and makes him a focal point for the actions of the novel. For example, both of the other main characters, Lula and Vitorino, pass by his house with great frequency. How far removed is this dusty road from the halls of Harvard where Quentin's portion of *SAF* begins—although Quentin feels himself as marginalized in New England as does Amaro in Pilar. In opposition to Amaro, whose thoughts never leave the confines of his immediate situation, Quentin's thoughts never seem to touch the present at all, revolving instead in a combination of memory and abstract philosophy, as his mental maps impose the urban Northeast onto the rural South. Yet, both characters end up taking their own lives, and the reader is meant to understand (or not) this event through each character's internal monologue. Amaro's and Quentin's chapters alternate between direct discourse, the gossip of society, and internal monologues revealing a prelinguistic semiotic. Whereas Benjy and Lula never incorporate social discourse, and Jason and Vitorino use it exclusively, Quentin's and Amaro's monologues oppose the social to the private, or in Kristevan terms, (Benjy's) Body of the Mother to (Jason's) Law of the Father, finding no rapprochement between them. As Marsha Warren has shown, Quentin, as speaking subject, "attempts to (re)create himself by rejecting the Law of the Father (time) as the regulating agency of discourse, and entering the Body of the Mother (space), thereby recovering the maternal semiotic flux" (101). Like Quentin, Amaro experiences this disruption, but like Jason he ruthlessly represses the semiotic through violence against his mentally disturbed daughter and his wife.

The leitmotif in Amaro's chapter is the word "lobishomem" ("werewolf"), a designation the townspeople bestow on Amaro to denote his liminal status. The yellow, sickly Amaro is neither impoverished nor wealthy, lives neither in town nor in the country (he lives at the edge of Lula's plantation), is neither wholly legal (he killed a man and had to flee to his present home, and he supports the activities of the cangaceiro Antônio Silvino) nor an outlaw. The name "Amaro" has associations both with "amargo" ("bitter"), which is his psychic condition, and with "amarelo" ("yellow"), the color of his skin. Stemming from a disease, the color yellow also functions symbolically as yet another in-between. Amaro's marginal status, not a slave and yet subservient to his landlord as though he were one, represents the real historical condition of the *agregado* in Brazil. Roberto Schwartz defines [negatively] the *agregado* as "neither proprietor, nor proletarian . . . [whose] access to social life and its benefits depended, in one way or another, on the favor of a man of wealth and power" (22). Since the "man of wealth and power" of this text, Lula, is himself lost to the world, Amaro's situation is desperate indeed. Amaro is, however, neither lost in an interior world, as is Lula, nor devoted to the exterior and social, as is Vitorino. Amaro's discourse, as Mercedes Risso points out, shows his interior conflict through the use of indirect versus direct discourse: "while direct discourse concretizes contact with external reality . . . free indirect discourse confronts us with interior, psychological reality cogitated privately by José Amaro" (29-30). Every time Amaro uses the word "lobishomem," it comes not from within himself, but as an echo from outside. Amaro's alternative to this social oppression is, as one might expect, nature. In the following passage, we can note the distinction between Amaro's identification with nature, and the constant references to language: "An enormous red moon appeared above the forest canopy on the ridge. His life, his whole life, was there. To think about it, his *compadre* Vitorino was happier than himself: Vitorino didn't suffer as he was now suffering; he didn't have that cold spot in his heart; that bitter mouth; that will, born of desperation, to do something he couldn't conceive of. . . . His wife approached to tell him about their daughter. His daughter's madness hurt him deeply. He felt an urge to order her to be quiet, to leave him in his terrible pain" ("Uma lua enorme, vermelha, aparecia por cima da capoeira do alto. Ali estava a sua vida, toda a sua vida. Bem reparando, mais feliz do que ele era o seu compadre Vitorino. Não sofria assim como ele estava sofrendo, não tinha aquele frio no coração, aquela boca amarga, aquela vontade de desespero, de fazer o que não sabia o que era. . . . A mulher chegou-se para mais perto para lhe falar da filha. Aquilo lhe doía fundo.

Teve ímpeto de mandar que se calasse, que o deixasse na sua dor terrível"; 50-51). We switch from the discourse of the narrator in the first sentence to the free indirect reproduction of Amaro's thoughts in the second sentence. The second sentence declares, or seems to declare, that Amaro finds his life in nature, although the exact referent of the deictic "ali" is not specified. It is "anywhere out of the world," we might say, in the "terra incognita" of the other side of the rise. As she is of Benjy's, Caddy is also the "object" of Quentin's discourse. His interior monologue plays out scenes of sexual interest and even of incest (undoubtedly never realized) that can be interpreted as an attempt to preserve family territory by preventing exogamy. In the first pages of Quentin's chapter, the grammar of English crashes upon the reef of the sister. After long and complex reproductions of Mr. Compson's discourse, the sister intervenes: "Like Father said down the long and lonely light-rays you might see Jesus walking, like. And the good Saint Francis that said Little Sister Death, that never had a sister. . . . Father said that. That Christ was not crucified: he was worn away by a minute clicking of little wheels. That had no sister" (86-87). Faulkner here reproduces at the level of language what can happen just as well in images: a train of thought derails and falls into the black hole of a single, overriding, anxiety-producing utterance. Here the image is that of the sister, which as we might expect summarizes for Quentin the disjunction between private and public discourse and behavior, between the Body of the Mother and the Law of the Father. In both sentences, the relative "that" has uncertain antecedents: in the first, either Francis or Little Sister Death could be sisterless; in the second case, "wheels" would be the grammatical antecedent, which the reader probably rejects on the basis of lexical absurdity. Another example from later in the chapter proceeds along the same lines. Here Quentin's semiotic fragments pepper the remembered discourse of his mother as she is riding in the car and talking to Herbert, Caddy's fiancé: "Why shouldn't you I want my boys to be more than friends yes Candace and Quentin more than friends *Father I have committed* what a pity you had no brother or sister *No sister no sister had no sister* Dont ask Quentin he and Mr Compson both feel a little insulted when I am strong enough to come down to the table I am going on nerve now I'll pay for it after it's all over and you have taken my little daughter away from me *My little sister had no. If I could say Mother. Mother*" (108). Again, the syntax of language is destroyed by the image of the sister. If the syntax of language is destroyed—as it will be progressively throughout the chapter—then its ability to produce propositional statements (such as Jason's, discussed above) is seriously impaired.

While an Oedipal interpretation of Quentin's obsession with his sister as a mother figure seems obvious, Richard Godden points to its historical dimensions. First of all, Quentin desires incest in order to paradoxically "preserve" Caddy's virginity, in accordance with the "raising on a pedestal" of the female offspring of the planter class—an enforcement of endogamy typical of privileged social strata. Secondly, the virginity of white women and fear of their violation was a Freudian inversion-repression of the actual satyr-like behavior of many white patriarchs in their slave quarters. The central conflict of *Absalom, Absalom!* is the planter Thomas

Sutpen's refusal to recognize his own mulatto son, the offspring of such behavior. Real practices of miscegenation, based on white males' relatively unrestricted access to black women, were censored by symbolizing all sexuality as black. Thus, Caddy's promiscuity appears to Quentin to be that of "nigger women" (59), and Jason perceives Miss Quentin's sexual behavior as that of a "nigger wench" (119). As Godden points out, Quentin's symbolic associations lead him to perceive Caddy's lovers as black as well, as expressed in the constantly recurring term "blackguards." Finally, Quentin's own desire for his sister must fall prey to the same racial symbolization, so that "incest . . . will shore up the integrity of his family and class, but can do so only in blackface, since white potency is inextricable from black forms" (108). Quentin's own whiteness is thus darkened by the supposed blackness of sexual desire, making him, like José Amaro, "yellow" ("yellow" was a term often used in the US South for light-skinned mulattos).

The other leitmotif in Quentin's monologue, then, is the "Negro." Both Quentin and José Amaro exhibit the exaggerated notions of slavery and freedom. Both define their whiteness over against a concept of slavery. At least one of the meanings for Quentin's shadow, which occupies him throughout his entire final day on earth, is that of the black who accompanies the white, giving his life meaning, as Luster accompanies Benjy. Eric Sundquist's claim that the Quentin Compson of *SAF* can only be adequately interpreted through the Quentin Compson of the later *Absalom, Absalom!* (1936) applies here. In that novel, Quentin unravels the mystery of the downfall of the white planter Thomas Sutpen by discovering that he had a mulatto son, Charles Bon, whom he repudiated because of his color and illegitimacy. Quentin's epiphany in *SAF* seems an abstract philosophizing upon this mythic event, before Faulkner had actually invented it:

> Quentin's realization that "nigger" is a form of behavior, an obverse reflection of the white people surrounding him . . . articulates two separate and parallel, but mutually dependent, phenomena: first, that the impenetrable mask of "Negro," however it divides the theories of historians and novelists, springs from a political reality that inevitably overleaps its varied social, physiological, and customary justifications; and second, that this masking is hallucinatory to the extent that, from a white perspective, it manifests the interiorization of racial trauma that led Faulkner, among others, to recognize that nigger—like all such epithets and possible in some instances no more or less than "Negro" or "black"—describes not a person but a projected image. (Sundquist 67)

Observing an old African American on a mule, Quentin is led to see in blacks "shabby and motionless patience, of static serenity: that blending of childlike and ready incompetence and paradoxical reliability that tends and protects them it loves out of all reason and robs them steadily and evades responsibility and obligations by means too barefaced and to be called subterfuge even and is taken in theft or evasion with only that frank and spontaneous admiration for the victor which a gentleman feels for anyone who beats him in fair contest, and withal a fond and unflagging tolerance for white folks' vagaries like that of a grandparent for unpredictable and troublesome children" (*SAF* 99-100). As Richard P. Adams's points out, this passage is a "collection of condescending clichés" (247). More interesting, however, is the fact that these clichés are derived from a number of different categories of life, so that they make Quentin a member of more than one class, and a resident of more than one place. Words like "shabby," "childlike," "incompetence," "evades responsibility," "theft," participate in Jason's racist fault-finding; words like "patience," "serenity," "love[s]," "victor," and "tolerance" take the high moral ground of patrician discourse. Historically speaking, then, Quentin's interior monologue bounces the xenophobic dirt-farmer's discourse off the landowning aristocrat's paternalism. These are not merely two philosophical positions, but two opposing platforms in Mississippi politics during Faulkner's formative years: Since Reconstruction, the "ignorant" and the "educated" of Mississippi's white electorate had been locked in a deadly struggle with blacks as the scapegoats, and now the "ignorant" were winning. For Faulkner . . . that was important, for the Faulkner family's way of thinking about itself had emerged from the rough and tumble of Mississippi politics. When Faulkner's great-grandfather William, the "Old Colonel," had been assassinated by his former business partner before he could serve in the state legislature of 1889, he had left the Falkner family with the question of "whether they were Cavaliers or rednecks, and not knowing would affect all of them profoundly" (Walter Taylor 7). The Compson family, guided by the mental maps of plantation life, also does not know, for they are both.

Just as Quentin's mental and verbal gymnastics most resemble Faulkner's own, so too we can discern a similarity of purpose between character's and author's vision of the black. We notice that Quentin's ultimate interpretation of blacks is fictional: they are (from the white point of view) an object of interpretation rather than an interlocutor. And as an object they represent the fulfillment of everything lacking in the white world. While Quentin's sympathetic meditations seem diametrically opposed to Jason's racist diatribe, they share this quality of using blacks as supplements. Not only that, but they share this strategy with *SAF*'s narrator, whom we encounter at work in the book's fourth chapter.

The fourth dimension

Although the thrust of both novels is what we may call perspectivist, both authors recuperate the relativism of the fragmented discourse of their main characters through the fourth, "supplementary" discourse of the narrator, that breaks through the walls of solipsism and extends the knowable community of the text. While Faulkner's fourth chapter is about Dilsey, the African American servant of the Compsons, it is equally about the narrator and, by extension, about the "implied author" and the moral vision of the text as a whole. Lins do Rego does not supply a fourth chapter because, as we have seen, the discourse of his narrator intervenes in each chapter to ironize the monologues of his main characters. (Lacan also supplies a fourth vertex to the Oedipal triangle, called the locus of Name-of-the-Father in the Other.)

The presence or lack of a recuperative "fourth dimension" in the two works has interesting parallels in the two authors' biographies: Lins do Rego abandoned Paraíba and the Northeast for Rio de Janeiro, while Faulkner, although he experimented with life in Ontario, New York, Europe, and Hollywood, maintained his base in Oxford/Jefferson and consciously cultivated the life of a Southern gentleman. In an introduction to *SAF* written long after its initial publication, Faulkner later referred to Dilsey as "the future, to stand above the fallen ruins of the family like a ruined chimney, gaunt, patient, and indomitable" (414). One sees here contradictions analogous to those in Quentin's monologue—in what sense can a ruined chimney be considered indomitable? The image of the ruined chimney also occurs in Lins do Rego's palette of images. In his case it is a *bueiro*, the smokestack of the sugar mill that provides the most visible sign of the plantation's well-being. When the plantation is producing sugar, its *bueiro* will be intact, clean, and pouring forth smoke. When the plantation is *de fogo morto*, and no longer processing sugarcane into sugar, the smokestack will become dilapidated and covered with plants. Such ruined smokestacks identify the sites of former *engenhos* all over the Brazilian Northeast. Like Faulkner, Lins do Rego sought to incorporate into his fiction the alternatives to social and economic death that he had described, and like Faulkner he saw the black population as integral to that solution. *O moleque Ricardo*, for example, follows the career of a black from a plantation neighboring Lula's as he flees rural peonage to Recife and becomes part of the urban proletariat. As mentioned, Zé Marreira takes over the Santa Fé. *FM* itself, however, provides no such alternative, either ideologically or in terms of its narrative focus.

The "plot" of the fourth chapter is simple enough: Dilsey goes with her family and Benjy to Easter church service and then returns. Jason discovers that his cashbox has been emptied out (perhaps an allusion to the Easter motif of Christ's empty tomb) and vainly pursues the culprit, his niece Quentin. Dilsey is unmistakably the center of this chapter. The remarks at the end of the Quentin section of this study give one reason why Faulkner chose third-person narration, which is that Dilsey as the representative of supposedly black values, such as "endurance," "simplicity," "faith," and "love," is a phantasmagoria. This is of course not to say that African Americans cannot have these values, but only that the essentialist reading of them into black culture as a whole—indeed, the idea that there is a black culture as a whole—is a symbolic gesture incapable of framing a reality that is diverse and contradictory. To give but one example of this, although Faulkner's portrait of Dilsey is sympathetic, we may note that she and her family are no less stuck in a time trap than are the Compsons. Thadious M. Davis has pointed out that the Compson and the Gibson family structures are almost exact replicas of each other: both families have deceased patriarchs, daughters who have departed, and the same number of children (72-75). In both authors, the narrative voice supplies not a coherent vision of reality that could transcend the perspectivism of the text as a whole; rather, the narrative voice confines itself to the critical function. As Eduardo Coutinho has put it, "the social or socio-existential problematic . . . achieves its critical dimension through the difference established between [the characters'] diverse voices and the narrator's discourse. The latter does not take up positions or point out paths; rather, by contrasting in a non-hierarchical manner with the voices referred to, it denounces their ideology and relativizes the cosmology presented" (439-40).

Dead father, absent mother

So far, I have been drawing an analogy between the works of two American regionalists. My reading of the fragmented structure of the two works as symbolic of the fragmented ego and disrupted communication process is at odds with the standard view of "psychological man" as modern and urban. The supposed symptoms of modernization and urbanization appear most strongly precisely in those contexts that have resisted these processes. While one cannot rule out as a source of these symptoms the superimposition these authors might have made of their urban experience on rural contexts, I would argue more directly that they are produced by the ruins of the plantation system. The solipsism reflected in the works' fragmented structures is the legacy of the plantation regime, whose "internal structure of ownership by Europeans, coerced labor by African or black slaves, and artisan or managerial roles filled by poorer whites, freed blacks, and persons of color" impressed itself upon the social hierarchy of an entire society (Schwartz, *Sugar Plantations* 251). Although nothing in Vitorino's background or behavior link him with the phantom middle class, he represents the synthesis of master and serf it was to provide in capitalist economies. The guilt-laden patrician Quentin also serves as the missing link between Benjy's and Jason's solipsisms, but like Amaro he fails to turn his thoughts into actions—his only action is the unthought, Little Sister Death. Thanks to his naiveté, Vitorino is able to achieve a masochistic freedom. The physical punishment that Vitorino suffers throughout the novel—beatings, rocks thrown by children, a whipping by the police—act as a symbol of the terrible price the middle-class paid in order to overcome patriarchy and achieve its role in history. Used loosely to indicate any system where the male enjoys advantages over the female, patriarchy in the narrower sense indicates a sociopolitical order in which there is a theoretically complete subjection or domination, usually embodied in both legal and religious practice, of all other members of the (usually extended) family by a single male, the Father. Patriarchy in the Americas was intimately bound up with slavery and colonization. As Charles Wagley has argued, "Brazil was founded as a society formed of two distinct castes—namely, a caste of European masters and a caste of Indian or Negro slaves" (142-43). This socio-political system was immediately modified when the masters took Indian or black women as wives or concubines, and thereby created intermediary castes of mixed-blood freemen. These intermediate groups, however, were nothing like a middle class, and suffered in their very origin from a lack of legitimacy. Amaro's yellowness, given in his name and his physical description, betrays his illegitimate, in-between status, which is also reflected in his ambiguous, landless "freedom." Faulkner, of course, has also dealt extensively with the problem of legitimacy, most notably in *Go Down Moses*, where Ike, the legitimate white inheritor of the McCaslin line, inherits the task of tracking down the many illegitimate progeny and delivering their compromised portion—a pay-off rather than an inheritance. Not surprisingly, Ike learns from all this to question and ultimately repudiate both his inheritance and fatherhood—he dies childless.

"The Carvalho family had exhausted itself in the literary dos Anjos family" (Lins do Rego, "Augusto" 16). While plantation society was in vigor—while the Father still lived—writers, decadent offspring incapable of inheriting the Father's place, were deprived of the *logos*, and unable to describe the system realistically. In the case of Brazil, Gilberto Freyre has detailed in the aptly titled "O Pai e o Filho" ("Father and Son") both the life-and-death power of the engenho *paterfamilias*, and the only alternative society provided for plantation scions: the school run by priests, where the emphasis was on universalizing and urbane conformity, discipline, order, and on European literary genres ill suited to convey American realities. The Brazilian novel thus begins not with realism but with exoticism, as José de Alencar writes not of the plantation, but of aboriginal alternatives to it. The most extensive fictions of plantation life, such as John Pendleton Kennedy's *Swallow Barn* (1832), are imbued with an ideology of benevolent paternalism. Literature seems capable only of depicting imaginary places, whether these be the Eden of nonagricultural Brazil, the plantation of happy slaves and benevolent masters, or the ghost of plantations past. After economic dominance passed to other productive systems, only the Father's ghost remained. Indeed, as Freud has pointed out in *Totem and Taboo*, the only real Father is a dead Father: "Why? Because the father whose 'power is derived from strength alone . . . does not hold any properly paternal authority.' For the Father to possess legitimacy as principle of authority, he cannot be the 'dominating and jealous male of the Darwinian tribe.' Once he is overthrown, murdered, and devoured, however, he can become the 'Father' in the pre-eminent sense required—that is, in the sense in which he will found the Law within the sons" (qtd. in Porter 101). Lula, Amaro, and Quentin all experience their fathers as powerful images, verbal or visual, in these texts. Not the Father himself, but the discourse of the Father constructed around this absent center holds these characters in its sway.

The main characters in these two texts are certain of their origins, and that certainty, a component of regionalist identity, cripples their every attempt at action. Haunted by a double absence, regionalist writers may now seize the *logos*, but only if they embrace with it the status of banishment: "Against the modifying whole of the father's Death, one chooses banishment toward the part constituting a fallen object or an object of love. . . . Without banishment, there is no possible release from the grip of paternal Death. This act of loving and its incumbent writing spring from the Death of the Father—from the Death of the third person" (Kristeva 391). Quentin and José Amaro are the symbols of banishment in the two texts, and as such they function as sacrificial offerings by these authors, who simultaneously lay claim to and repudiate their plantation places.

Chapter Six

About Roads and Rivers and Arguedas's Ethnogeographies

Something curious happens in the following passage from the Peruvian José María Arguedas's (1911-1997) apology for how he came to be a writer: "I began to write when I read the first narratives about Indians, which described them in such a false way. . . . In those stories the Indian was so disfigured, and the landscape so silly and saccharine or so foreign that I said: 'No, I must write this as it is, because I have enjoyed it, I have suffered it'" ("Yo comencé a escribir cuando leí las primeras narraciones sobre los indios, los describían de una forma tan falsa. . . . En estos relatos estaba tan desfigurado el indio y tan meloso y tonto el paisaje o tan extraño que dije: 'No, yo lo tengo que escribir tal cual es, porque yo lo he gozado, yo lo he sufrido'" (Arguedas, :"Yo soy hechura"; 251). The Indian and the Peruvian landscape condense into a single object, "lo" ["it"], which the writer has enjoyed and suffered. Grammatically speaking, it would have been possible to keep them separate, using "los" and "tales" for the Indians, but Arguedas chooses to merge them. Judging from his narratives, from the stories of *Agua* (1935; "*Water"*) to the novels *Yawar fiesta* (1948; "*Festival of Blood"*) and *Los ríos profundos* (1958; "*Deep Rivers"*), Indian and landscape were indeed inseparable, like the stones of the Inca wall that had been laid together so carefully and have now lain together for so long that not even a knifeblade can pass between them. Similar intimacy obtained between Arguedas's scientific, artistic, and political writing projects. Arguedas did not strive to create finely crafted stories and novels that could stand by themselves as aesthetic objects. Many critics have noted the lack of emplotment in his fiction, and the predominance of description over action—a relatively inactive diataxis, we might say, that is always returning to the mimetic plane. His fiction stands as a primary example of the archival impulse that founds the Latin American novel, as Roberto González Echevarría has demonstrated. So closely related were Arguedas's artistic and "scientific" intentions that the same true imaginary places appear in both. (Those who have written of the interaction between these domains include: Petra Iraides Cruz Leal and Rita de Grandis.)

So strong was this author's sense of place that even his optimism over the election of Peruvian President Fernando Belaúnde Terry, who had run on a platform of agrarian reform, was expressed in geographic terms. Terry would provide "a firm, dynamic, modern government, thoroughly familiar with our territory ['conocedor a fondo de nuestro territorio'] and with the enormous problems it faces. [Terry] has stated that he appreciates and loves the disinherited sector of the Peruvian people that, until recently, was despised and tormented as if such behavior belonged to the natural order of things" ("La cultura" 6). Once again we see a merging of people and action into a landscape, the "natural order of things." The "disinherited sector" are the Indians and, in a larger sense, all those who live far from the coastal centers of power and wealth. Like Euclides da Cunha and Juan Benet, then, Arguedas structures his work around the binary opposition between center and region, which as we will see finds dramatic physical expression in the landscapes of Peru. However, Arguedas reverses the "terra ignota" theme found in those authors. In his work, landscape is known too well, becoming an animate being that exacts sacrifices and preserves a way of life often detrimental to those who live on it. Fernando Alegría has described the sudden "standing up" of Arguedas's work in words reminiscent of Edouard Glissant's description of Faulkner: "All of a sudden, this writing that could be a catalog of churches, plazas, walls, handicraft and ruins, begins to live independently: the stones speak; the patios tremble" (275).

One might apply to Arguedas's "Peru" what Régis Durand has said of Faulkner's Yoknapatawpha: "The [Arguedian] territory is above all topological; that is to say, it is not given as a complete entity from the beginning. The space is constructed through a series of steps, successive investigations, onward movements and traverses. These carry the name of the author's fictional works" (n.p.). The various locations within this territory, as well as the boundaries that separate them (symbolized as deep rivers), the landscape modifications that unite them (roads and bridges), and the processes of transculturation that deterritorialize their inhabitants, together create a complex ethnogeography of Peru. Arguedas's unique achievements as a writer have given him a prominent place in Latin American letters, and an extensive body of criticism on his work by Sara Castro-Klarén, Antonio Cornejo Polar, Ángel Rama, and Mario Vargas Llosa, to name only the most prominent among many. Although this criticism often devotes itself to the ethnographic impulse that unifies Arguedas's generically disparate writings, the relatively narrow palette of geographic features he invests with ideological meanings has not been assembled in one place before.

Laws of altitude

Arguedas's geographic metaphors are dynamic and constantly transmogrifying. Arguedas views landscape as process rather than as object. In this sense, the particular kind of transculturation he documents can be described as deterritorialization and reterritorialization. This term is of central importance to several works of Félix Guattari and Gilles Deleuze, particularly to their study of Franz Kafka and to *A Thousand*

Plateaus. Yet, in line with their philosophy that seeks to break with static traditions of fixed meaning and eternal truths, they never give a full definition of the term. On the psychic level, these terms serve to replace static Freudian terminology for ego formation: the infant is deterritorialized from the breast and reterritorialized into language. Yet the terms have also been applied in their literal sense, to describe cultural changes stemming from uprooting and migration. This severing of the body from the land produces deterritorialization: "Deterritorialization is the movement by which 'one' leaves the territory" (Deleuze and Guattari 508). The uprooting of individuals divests their bodies of identity. They are then characterized by mobility, changeability, and a transitory nature, thereby resisting hegemonic control. Arguedas's work takes full cognizance of these processes, but views them with ambivalence. The Indians are the most territorialized of Peru's social groups, completely tied to a landscape they view as animate and endowed with a fate linked to their own. This feeling is shared by many of Arguedas's protagonists, and treated sympathetically. At the same time, Arguedas shows in his narratives that only deterritorialization can free the Indians from their domination by elites.

Peru consists of three distinct topographies: a dry coastal plain, the Andes, and the Amazon Basin. Arguedas hardly ever concerns himself with the last of these, which is the least populated. A major exception is the character of Esteban presented in chapter four of *El zorro de arriba y el zorro de abajo* (1971; "*The Fox from Above and the Fox from Below*"), an Indian from Parobamba who has lived and worked in all three regions. For the first two, a simple relationship can be charted over the extent of the fiction (see Figure 18).

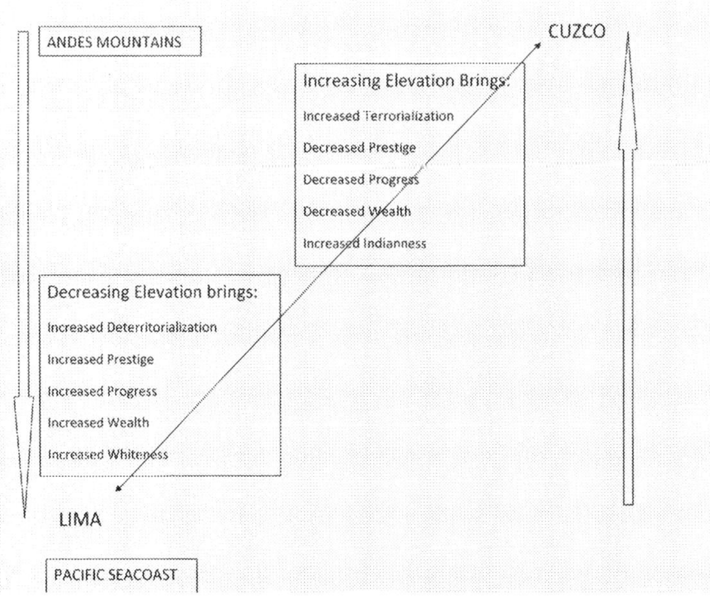

Figure 18. Altitudinal binarisms in the work of Arguedas

This binary scheme appears with great clarity in the early work of Arguedas, while the later fiction, such as *El zorro*, shows its complication through the deterritorializing processes of international capitalism and globalization (this despite the fact—and the dialectic is typical of Arguedas—that the title of the novel expresses the binarism in the clearest fashion imaginable). In terms of the development of Arguedas's fiction, from the early stories to this last novel, we could speak of a gradual recognition of the Indianness of the coast and of a descent into its environs. Inasmuch as his fiction works painfully towards a sublimation of the binarisms of the mental map sketched above, we might speak of Arguedas as a "Piruvian" nationalist. That is, Arguedas's writing attempted to redefine the territory of Peru outside the vision of its dominant hierarchies, incorporating Indian visions of reality the way his narratives incorporate Quechua into Spanish, resulting in ungrammaticalities, new vocabulary, and mispronunciations such as the one reflected in the new spelling of "Peru."

Arguedas's Law of Altitude of Peru is that decreasing elevation brings increasing prestige, progress, and economic independence. Increasing elevation "Indianizes," but it also brings a certain degree of independence from dominance. Millions of highland Indians have migrated to coastal cities such as Lima, Chimbote, and Ica, and Arguedas is fascinated by their transculturation within a new environment. To give one example, the Tinki community lives higher than that of San Juan, and this difference in elevation has consequences. Their health suffers from the high altitude and lack of nutrition, but they have an air of authenticity and freedom, at least to the young narrator of Arguedas's early story, "Agua": "I observed the faces of the comuneros, one after another: all of them were ugly, with yellow eyes. Their skin was dirty and burned by the cold, their hair long and greasy; almost all of them looked broken. . . . But they maintain a better expression than did those of San Juan, they did not look cast down, they talked and laughed with Pantaléon in loud voices" ("Yo miré una a una las caras de los comuneros: todos eran feos, sus ojos eran amarillosos, su piel sucia y quemada por el frío, el cabello largo y sudoso; casi todos eran rotosos. . . . Pero tenían mejor expresión que los sanjuanes, no parecían muy abatidos, conversaban en voz alta con Pantaleón y se reían"; 24-25). This mixture of repulsion at the physical features and hygiene of "true" Indians and admiration of their preservation of independence remains a constant in Arguedas's work. Arguedas's own experience provided much of the data for his mimesis. Born to an itinerant country lawyer in 1911, Arguedas lost his mother when he was three, and he was left in the care of his paternal grandmother in Andahuaylas, in the southern mountains of Peru. Arguedas learned to know the "Indian town" of Puquio, scene of *Yawar fiesta*, when his father married a rich widow who forced her step-son to live with the Indian servants. By 1923, Arguedas was accompanying his father on his trips through the Andes before attending a religious school in Abancay. Those travels, and the sojourn in that town, form the basis of *Los ríos profundos* (*"Deep Rivers"*). Simultaneously with the beginning of his postsecondary studies in Lima, Arguedas began publishing stories in Spanish and translations from Quechua. His first collection of stories, *Agua*, gained him international notoriety by winning second prize in a contest sponsored by the

Revista Americana of Buenos Aires. The title story of this collection develops many of the themes, ideas, and spaces found throughout Arguedas's Andean fiction. The story is set in San Juan, but the social hierarchies and conflicts will be revisited in the Puquio of *Yawar fiesta* and *Todas las sangres* (1964). The Indians who still live a traditional life, called *comuneros*, are controlled by the dominant class, called *mistis*. "Misti" is a Quechua term that simply means "white"; however, in reality it is applied to landowners as distinct from peons, regardless of race. Similarly, the designation "Indio" refers not to genetic inheritance, but to language, dress, and solidarity with a community defined by place. Ernesto, prototype for the hero of *Deep Rivers*, accompanies the horn player Pantaleón. As always in Arguedas's early fiction, the first-person narrator of this story is a version of Arguedas himself. The conflict of the story centers around water, which is controlled by a single *misti*, Don Braulio. In a period of drought, he insists on keeping all water for the landowners, and giving none to the Indian communes. Economic ruin foments rebellion, led by the Indians Pantaleón, Don Wallpa, and Don Pascual. Pantaleón, the instigator of the rebellion, has been rendered articulate, transculturated by his experiences at the coast. When Don Pascual begins the ceremony of dividing the water, and gives it to the needy, Don Braulio brings out his revolver and begins firing, eventually killing Pantaleón.

Arguedas saw his task as the rewriting of a world view expressed in Quechua into the Spanish language, thus giving some degree of balance to the process of transculturation. A relatively simple example of this process occurs when the Spanish word "sol" ("sun") does not appear in the story, although the sun is an important agent of the action, causing and aggravating the drought. Instead, the phrase "Tayta Inti" is used, designating an animate being, just as authors based in classical mythology might circumlocute with "Phoebus Apollo." In "Agua," as the comuneros of Tinqui descend from the heights to San Juan, they describe the mountains, Viseca and Ak'ola. The narrator remarks that the Indians of San Juan see the two mountains as rivals, who descend at night to bathe in the Viseca. The Tinkis also speak of these peaks as if they were human:

> [Mount] Viseca shouts louder. That's right! Viseca is who proclaims father. Tayta Chitulla is his master; he's the Kanrara of Ak'ola just. Kanrara? Tayta Kanrara wins out over Chitulla, he's wilder. True. His head is pointed like Don Cordova's spiked fence. And Chitulla. Four Kanraras could fit in his belly. The Indians looked from one peak to another. They compared them with all seriousness, as if they were looking at two men. . . . The Indians of San Juan say that the two peaks are rivals and that in the dark of nighttime, they go down to the banks of the Viseca and dive in, from shore to shore.
>
> Viseca grita más fuerte.—¡Claro pues! Viseca es quebrada padre; el tayta Chitulla es su patrón; de Ak'ola es Kanrara nomás.—¿Kanrara? Tayta Kanrara le gana a Chitulla, más rabioso es. —Verdad. Punta es su cabeza, como rejón de don Cordova.—¿Y Chitulla? A su barriga segura entran cuatro Kanraras. Los indios miraban a uno y otro cerro. Los comparaban, serios, como si estuvieron viendo a dos hombres. ... Los indios sanjuanes dicen que los dos cerros son rivales y que en las noches oscuras, bajan hasta la ribera del Viseca y se hondean ahí, de orilla a orilla. ("Agua" 24-25)

Quechua words such as *tayta* and *kanrara*, together with a simplified Spanish grammar, help create a language in which indigenous thought is expressed in a Western idiom.

In the story, "Los Escoleros," the cliff named Jatunrumi is also treated by the inhabitants of Ak'ola as if it were alive: "Jatunrumi is the largest cliff in Ak'ola. It sits at the side of the road that leads to the *puna*s, fixed to the mountainside. From the side of the road it doesn't look so high, but seen from the pond that carries its name, it seems to be a peak; you get dizzy if you look at it for a long while. To climb to Jatunrumi's head was a feat attempted by the older schoolboys" ("Jatunrumi es la piedra más grande de Ak'ola, está sentada a la orilla del camino que va a las punas, clavada en la ladera. Por el lado del camino no se le ve tan alta, pero mirada desde el potrero que lleva su nombre, por la parte baja de la ladera, parece un cerro, da vueltas la cabeza cuando se le contempla largo rato. Subir hasta la cabeza de Jatunrumi era proeza de los escoleros mayores y más valientes"; 64-65). When Juan climbs the rock, and then can't get down, he considers that the cliff has captured him: "Go down? Never! Jatunrumi wanted me for himself, surely because I was an orphan; he wanted to make me stay forever in his top. Like a sparrow fallen into a trap, I was going around and around on his top without finding a way out" ("¿Bajar? ¡Nunca! Jatunrumi me quería para él, seguro porque era huérfano; quería hacerme quedar para siempre en su cumbre. Como el gorrión que ha caído en la trampa, daba vueltas en la cumbre de la piedra sin encontrar camino"; 65-66). This is not just the fantasy of a young child, but rather his recollection of the stories told by the Indians concerning all the peaks of the region: "It is said that from time to time, these peaks get hungry and grab a *mak'tillo*; they eat him up and keep him in their insides. Sometimes the *mak'tillo* prisoners remember the earth, their people, their mothers, and they sing sadly. Have you never hear Jatunrumi sing thus? Anyone's heart would cry when in the dark nights, after the rain has stopped, for example, Jaturumi sings in a sad, thin voice. But it is not Jatunrumi, but the voice of all the poor *mak'tillos* he has captured" ("De tiempo en tiempo, dice, sienten hambre y se llevan a un mak'tillo; se lo comen enterito y lo guardan en su adentro. A veces los mak'tillos presos recuerdan la tierra, sus pueblos, sus madres y cantan triste. ¿No le has oído tú cantar a Jatunrumi? El corazón de cualquiera llora si en las noches negras, cuando ha parado la lluvia, por ejemplo, canta Jatunrumi con voz triste y delgadita. Pero no es la voz de Jatunrumi, es la voz de los probres mak'tillos que se ha llevado"; 66). Symbolically, Juan and the *mak'tillo*s' imprisonment represents the absolute territorizalization of the Indians, their rootedness in a magical landscape. Arguedas shows the high price paid for deterritorialization. Juan is finally freed from the rock by the foreman, Don Jesús, who then proceeds to beat him soundly. The reader is shocked when the story suddenly turns from the rescue of Juan to his cruel beating with a whip. Don Jesús enforces the rule of the evil Don Ciprián, a *misti* who controls the whole community, and who precipitates the story's catastrophe when he appropriates and then shoots a favorite milk cow belonging to an Indian family.

In the second chapter of *Yawar fiesta*, Arguedas describes in detail the geography of economic stratification in the Andes. Basically, the *mistis* have taken the valley land for themselves, and the Indians have been forced ever higher, onto the *puna*. The difference between the two areas is shown not only in natural signs, such as the absence of trees. Absent as well are the fences and other boundaries between personal possessions: "The puna really belonged to the Indians: the puna with its animals, with its grass, with its cold winds, and its showers of rain. The mistis feared the puna and let the Indians live there. . . . Each of the Puquio communities ["Cada ayllu de Puquio"] had its own grazing lands. They were the only boundaries on the punas: a brook, the brow of a hill marked the claims of each ayllu" (78; "De verdad la puna era de los indios; la puna, con sus animales, con su pastos, con sus vientos fríos y sus aguaceros. Los mistis le tenían miedo a la puna, y dejaban vivir allí a los indios. . . . Cada ayllu de Puquio tenía sus echaderos. Ésa era la única división que había en las punas: un riachuelo, la ceja de una montaña, señalaba las pertinencias de cada ayllu"; 10). Eventually, the immense profit to be had from the sale of beef to urban centers leads the *mistis* to dispossess the Indians of even this last, inhospitable environment. Their forced descent from the *puna* is fatal: "And in Puquio there would be one more hand to work the prominent citizens' field crops, or to be 'hooked' to go to Nazca or Acari, to work on the coast. There they would make good bait for the malarial mosquitoes. . . . On their way home, they would find eternal rest in the sun-scorched sands, on the slopes, or up on the puna" (*Yawar Fiesta* 16; "Y en Puquio había un jornalero más para las chacras de los principales, o para 'engancharse' e ir a Nazca o Acarí, a trabajar en la costa. Allá servían de alimento a los zancudos de la terciana. . . . A la vuelta, 'cansaban' para siempre en los arenales caldeados de sol, en las cuestas, en la puna "; *Yawar fiesta* 82). What little diegesis *Yawar fiesta* develops concerns the capturing of the wild bull of the puna, Misitu, and his sacrifice in the traditional Indian bullfight. Clearly, Misitu represents the remnants of Indian spirit and world view that must eventually be given up, but not without a fight.

In the story, "Agua," since we have seen the *tinkis* descending from the *puna* to the town controlled by Don Braulio and other *mistis*, we associate the elevation they live at both with their decrepit physical condition and with their independence. Outside of his fiction, Arguedas wrote often of this divide. Note how the following discussion opens with a reference to its overcoming through roadways: "Before these new pathways opened, there existed a substantial disjunction between the classes in power ["classes dirigentes"] in the Capital [Lima] and in a few important cities of the coast and the Andean region. The Andes constituted, especially for residents of Lima, an unknown region, inhabited by the inferior and even the abject. The attitude of the upper classes of Lima had been formed by the colonial tradition and European models. . . . This attitude was rigid: an impenetrable wall kept the middle and elite classes from comprehending even *mestizo* culture, not to mention Indian, and from comprehending the behavior of the classes in power in the Andean towns. Peru was really two countries in one" ("Sabogal" 241). As in the Brazil of da Cunha, in Peru the Europeans cling to the coast and the mountains belong to the Indians and a few *mistis* (the other main region of Peru, the Amazonian Basin, is too sparsely populated to complicate this binary cultural dynamic). In compensation, however, in an ideological movement familiar to us from the case of da Cunha and the *sertão*, the Andean region came to be identified as the seat of "authentic" Peruvian culture. A decisive moment in this creation of region and regionalism can be found in José Carlos Mariátegui's *Siete ensayos de interpretación de la realidad peruana* (*"Seven Essays on the Interpretation of Peruvian Reality"*), published in 1928 when Arguedas was seventeen years old. Mariátegui does not hesitate to identify the cultural center of Peru as the departments of Cuzco, Arequipa, Puno, and Apurímac—precisely the territory Arguedas knew best at that time: "These departments constitute the best-defined and most organic of our regions. Among these departments the exchange and linkage maintain an ancient unity: that inherited from the time of Inca civilization. In the south, 'region' rests solidly on the historic bedrock. The Andes are its bastions" (208). These bastions were created by a topographical fact: the flat coastal plain becomes very narrow in the south of the country, giving no room to the Spanish predilections for the lowlands. Hence, considering Peru as a whole, the opposition West-East also becomes one of North-South. Mariátegui's term "bastion" is carefully chosen, and points to the isolation caused by, among other things, the topographic features that make travel from coast to mountain extremely arduous.

Peruvian national reterritory

These simple geographical differences trace themselves onto the psyches of the characters in Arguedas's writings. Paradoxically, however, the characters who are most affected by the binarism of mutual isolation and misunderstanding are the ones who transcend it through travel. The narrator of "Orovilca," of Andean origins but attending school in Ica, on the Nazca plain, explains his predicament: "I was a first-year student who had recently arrived from the Andes, and I tried to be inconspicuous, because in those days, in Ica the same as in all coastal cities, indianized mountain people were looked down upon, not to mention those coming from small towns" (my translation; "Yo era alumno del primer año, un recién llegado de los Andes, y trataba de no llamar la atención hacia mí; porque entonces, en Ica, como en todas las ciudades de la costa, se menospreciaba a la gente de la sierra aindiada y mucho más a los que venían desde pequeños pueblos"; 76). Here we see process as a complete identification between geographical location and social position. The previous paragraphs have identified the narrator with the "chaucato" bird, who is "campesino" (75)—literally, "from the countryside," but also a word for "peasant" and by implication, "Indian." The narrator is not Indian, but "aindiada," indianized, because he has only recently arrived from the highlands, where all social classes undergo influence from native cultures—a process symbolized by the Indian bullfight or *yawar fiesta* in the novel of that name. With time, the patina of Andean culture might be worn away, both through contact with the Europeanized world of the coast, and through the maturation process that kills the young person's magical view of the world.

Such are some of the "deep rivers" that divide Peru. However, Arguedas saw the divide being overcome by the "nuevas vías" of the highway. Arguedas begins his homage to the indigenist plastic artist, José Sabogal, indirectly, with words about roads and their influence on Peruvian culture: "During the last few decades, the influence of modern culture has penetrated much more fully into the Andean regions of Peru, in consequence of the opening of pathways of mechanical communication" (241). And, "The Amauta movement coincides with the opening of the first highways" (242). Amauta was the name of the first serious Peruvian movement for indigenism in art, culture, and literature. Arguedas views the development of Peruvian culture in the most material terms possible. Natural barriers to communication enable the preservation of traditional culture. When these are overcome by mechanical means, a double-edged process of transculturation begins: traditional ways are displaced by modernity, but the neo-Europeans of the coast can also experience *mestizo* and Indian cultures firsthand, and learn to appreciate them. The neo-European Sabogal used the highways to penetrate the mountains, and the result was that he learned to put the Indians center-stage in some of his works. Arguedas continues with an anecdote of the production of ceramic bulls in Pucará. When the Pucarans began to produce for a coastal and tourist souvenir market, rather than for a local one, they at first changed their style, as if the bulls were objects of devotion in a religion which they knew Limeños did not practice. However, the nonlocals quickly learned to appreciate the older, more traditional bulls, and the Pucarans eventually returned to their original style. Local culture was at first deterritorialized by "nationalization," and then reterritorialized with a more solid material base than it had enjoyed before.

Arguedas shows a similar process in the pages of *Yawar fiesta*. In competition with Coracora, another Indian community, the comuneros of Puquio build a "street on big mountain" (60; "calle en cerro grande"; 120) that goes to Nazca in just twenty-eight days:

> The Lima newspapers mentioned the Nazca-Puquio highway. One hundred and eighty miles in twenty-eight days! By popular initiative, without government support! And that's when all the other towns began. In the north, middle, and in the south of the country, even in the jungle, people held large town meetings in the squares; they sent telegrams to the national government and began work on their own. Anyone at all would trace out the highway's course to the coast, figuring out how to cross the mountains and the valleys.... And it was by this same highway that the 2,000 Lucaninos and the people from Coracora got to Lima. At the same time, on all the new roads, highlanders from the north, south, and middle of the country went down to the capital. . . . Once again, after 600 years, perhaps after 1000 years, Andean people were going down to the coast in multitudes. (*Yawar Fiesta* 66-67)

> Los periódicos de Lima hablaron de la carretera Nazca-Puquio. ¡Trescientos kilómetros en veintiocho días! Por iniciativa popular, sin apoyo del Gobierno. Y desde entonces empezaron todos los pueblos. En el norte, en el centro, en el sur, hasta en la selva se reunían en las plazas de los pueblos,

> en cabildo grande; pasaban telegramas al Gobierno, y comenzaban el trabajo por su cuenta. Cualquiera hacía el trazo de la carretera a la costa, calculando los cerros y las quebradas. . . . Y por esa carretera llegaron a Lima los dosmil lucaninos, y los caracoreños. Al mismo tiempo, por todos los caminos nuevos, bajaron a la capital los serranos del norte, del sur y del centro. . . . Después de seiscientos anõs, acaso de mil años, otra vez la gente de los Andes bajaba en multitud a la costa. (*Yawar fiesta* 125-26)

The act of road-building parallels the bullfight as an assertion of Indian power and, at the same time, an undermining or "killing" of it—in each case, the landscape claims sacrificial deaths. In the supplement he wrote to the end of *Yawar fiesta*, Arguedas points out another effect of these roads, which is nothing less than the destruction of regional cultural autonomy: "The highways to the coast have broken up the Huamanga (Ayacucho) cultural area; Parinacochas and Lucanas provinces now belong to the growing area of influence of Ica and the coast" (151).

Transculturation of the native thus involves a farewell to the magic power of the earth. In a later novel, *Todas las sangres*, Arguedas contrasts the careers of two *misti* brothers, Don Bruno and Don Fermín. The latter becomes an entrepreneur who wishes to give Indians employment by opening a mine. Fermín expresses the necessity for such a transformation in absolute terms. His analysis could serve as an interpretation of the "Agua" story: "To believe that mountains suffer and have power turns the Indians into abject beings, because sensitivity is foremost with them, and the individual almost doesn't exist . . . Now they are barbarians who eat earth and cry too much" ("Creer que la montañas sufren y tienen poder los convierte [a los indios] in seres indeseables, porque la sensiblería ocupa en ellos el primer lugar, y el individuo casi no existe ... Ahora son bárbaros que comen tierra y lloran con exceso" ; 156). The insulting, hyperbolic remark that the Indians eat earth reinforces their deep connection to the landscape. Don Fermín's formulation makes the process of transculturation retroactive: the Indian's animist beliefs "convert" them, as though they had believed something else before becoming Indians. Their territorialized condition disqualifies them from the capitalist world, which requires senses of individualism and of dominance over nature. It is no accident that the name "Peru" also recurs frequently in Don Fermín's discourse, as in the following passage: "This country should be, can be great. Only capitalism will allow it to become great. . . . Peru is embarrassing: idol-worshiping Indians; illiterate, with a gentleness that is primitive and despicable, a people that speaks a language that cannot express rational thought, but only lamentation or carnal love" ("Este país merece ser grande, puede serlo. Únicamente el capitalismo lo conseguirá. . . . El Perú da vergüenza: indios idólatras; analfabetos, de ternura salvaje y despreciable, gente que habla una lengua que no sirve para expresar el raciocinio sino únicamente el llanto o el amor inferior"; 235). One of the keys to national identity is a common language, and Don Fermín rejects the possibility of Quechua adding its resources to Spanish, or sharing with it the status of a national language. Indeed, Spanish remains the sole national language of Peru today. (The only native language to become a national language in the Americas has been Guarani, which in Paraguay shares official status with Spanish.) Only through breaking the ties to their land and their language can these Indians be reterritorialized as Peruvians. Don Bruno, on the other hand, reterritorializes himself as an Indian by speaking Quechua and by acting in solidarity with the Indians. The narrator consistently describes Bruno in terms that make him an image of the landscape—for example, his eyes "brillaban cristalinamente" (my translation; "shone like crystals"; 39). "Crystalline," in fact, is a *leitmotif* in Arguedas, referring to the clear water of the deep rivers of the Andes, as in this passage from *Deep Rivers* where Ernesto compares himself to a river: "I must be like that clear, imperturbable river" (*Deep Rivers* 63; "Había de ser como ese río imperturbable y cristalino" *Los ríos profundos* 70). Unlike Don Fermín, Bruno accepts the trade-off of increased power for reduced liberty and individual choice. The novel's thesis revolves around this dilemma, embodied in the two brothers: "the comuneros suffer a restricted liberty because spiritually they are rooted in their own value system. On the other hand, in the great cities the Indian loses his only external source of strength: the landscape" (340).

About roads and rivers, mountain and ocean

Whether he names him Ernesto, or Juan, or leaves him unnamed, as in another early story, "Orovilca," Arguedas prefers the adolescent boy as a narrator. This narrator always lives in conflicted circumstances and presents a double consciousness. His mental map of the surroundings is mostly "Indian." He perceives every natural thing as alive, and as possessing a story that the narrator must unravel and tell to the unbelieving, Spanish world to which he is connected through family or formal education—the "internado" or boarding school, for example, is the scene of both "Orovilca" and also of *Deep Rivers*. At the end of the former short story, the hero Salcedo mysteriously disappears after winning a fight with the school bully, Wilster. The narrator clarifies it all to the school authorities: "They didn't find him. I told the principal that we should look for him in the path from Orovilca to the sea. Behind the groves of *huarangos*, among the evils that surround the lake, undulating waves of snakes are traced in the sand. These traces climb a bit on the desert slope. 'He walked here! Here's where he must have started his voyage to the sea!' They listened to me as if I were a delirious child, as if I a boy addicted to apparitions and fantasies, just like all those who live among the deep rivers and the immense mountains of the Andes" ("No lo encontraron. Yo le dije al Inspector que lo buscáramos en el camino de 'Orovilca' al mar. Detrás de los bosques de huarangos, entre las malezas que rodean la laguna, huellas ondulantes de víboras hay marcadas en la arena. Las huellas suben algo por la pendiente del desierto. ¡Por allí ha andado él; por ese punto debió iniciar su viaje al mar! Me escucharon como a un niño delirante, como a un muchacho adicto a las apariciones e invenciones, como todos los que viven entre los ríos profundos y las montañas inmensas de los Andes"; "Orovilca" 95-96). The last phrase is an extended metonym, where "Indio" is replaced by the description of nature in which the Indians normally find themselves. Ethnographic identity becomes topographic.

Simultaneously, the narrator seems to hide himself within the immensity of natural formations in the Andes. The deep rivers (as if Arguedas were making a forward allusion to his own novel, to be written several decades later) form an insurmountable border, across which communication is impossible. By the same token, the hero Salcedo undergoes the ultimate deterritorialization, leaving the striated spaces of mountains, roads, and rivers for the ultimate "smooth space" of the sea or of the cosmopolitan port towns. Deleuze and Guattari state: "The smooth spaces arising from the city are of a counterattack combining the smooth and holey and turning back against the town: sprawling, temporary, shifting shantytowns of nomads and cave dwellers, scrap metal and fabric, patchwork, to which the striations of money, work, or housing are no longer even relevant" (481). Such is Arguedas's description of Chimbote in *El zorro*.

Rivers symbolize the isolated nature of Andean life, which allows each community to operate under its own customs and laws, which nevertheless somehow all develop in analogous fashion, to the detriment of the indigenous inhabitants. On the Europeanized coast, the water disappears from the surface and goes underground: "In the bowels of the earth, in the pockets where perhaps only the roots of very old ficu trees reach, there is crystalline and fertile water. . . . The voice of the chaucato bird is the only sign that below the sun we have this deep current. . . . Why is it the chaucato who discovers the snake in the dust, who is the color of the dust and made of evil fire? Absolute opposition! The snake is spawned out of a special, rejected part of the dust, which has absorbed the rays of the sun, of the malignant part of the sun. The water negates the snake; it douses the heat!" ("El el fondo de la tierra, en los núcleos adonde quizá sólo llega la raíz de los ficus muy viejos, hay agua cristalina y fecunda. . . . La voz del chaucato es el único indicio que bajo el sol tenemos de esa honda corriente. . . . ¿Por qué el chaucato descubre en el polvo a la víbora, que es del color del polvo y hecha de fuego maligno? ¡La oposición absoluta! La víbora brota de una parte especial, negada, del polvo, que a su vez aprehende los rayos del sol, de la parte maligna del sol. ¡El agua la niega; apaga el ardor!"; "Orovilca" 77). Only the mysterious lagoons, for example, Orovilca, that the narrator visits with his friend Salcedo, provide a non-continuous, magical line of communication from the Andes to the sea, reminiscent of Juan Benet's hydrographic projects: "[The lake] appears isolated, as one of the earth's mysteries, because the Peruvian coast is a southern desert, in which the valleys are just thin threads linking the ocean with the Andes. And the soil of these oases produces more than any other in the Americas. It is dust that the water from the Andes has renewed every summer, for thousands of years" ("Aparece [la laguna] singularmente, como un misterio de la tierra; porque la costa peruana es un astral desierto donde los valles son apenas delgados hilos que comunican el mar con los Andes. Y la tierra de estos oasis produce más que ninguna otra de América. Es polvo que el agua de los Andes ha renovado durante milenios, cada verano"; 82-83). The language is almost that of a geography textbook, but infused with poetic interjections and a sense of mystery that derive partly from the discontinuity of the landscape. The Andes are at once distant and unapproachable, but also constantly present through the nutrients they provide the soil when the summer run-off swells the intermittent rivers of the coastal desert, as Salcedo describes it: "The water comes to Nazca in January; it comes slowly and the riverbed swells slowly, rises, until it forms that drag along roots dislodged from the bottom, and rocks that spin and crash in the torrent" ("Llega el agua en enero a Nazca, viene despacio y el cauce del río se hincha lentamente, se va levantando, hasta formar trombas que arrastran raíces arrancadas de lo profundo, y piedras que giran y chocan dentro de la corriente"; 86). The Quechua word for this phenomenon is "lloqlla," and we will see below how Arguedas returns to the image in *El zorro*. Significantly, the narrator then asks Salcedo whether he has visited the Andes, and he responds in the affirmative—"I got as high as 4200 meters" ("a 4200 metros"; 86). Like the movement of tides, or the hydrological cycle itself, the downward flow of water is balanced by the journey upward of a Nazqueño into the mountains. The comparison is of apples and oranges, unless one conceives—as an Indian would—of water as a living, intentional being, the equivalent of a person.

Roads, on the other hand, symbolize the overcoming of this isolation by technological means. The type of road that does this is a "carretera," for which we could also use the term "highway," but which the Quechua speaker in Yawar fiesta calls simply "mountain street." It is a major road, linking cities or provinces with each other. It is European in origin, and designed for automobile rather than for foot or mule traffic. The Incas used "senderos," such as the famed Inca Trail. Roads allow increased movements of goods and populations, with far-reaching effects, both negative and positive, on the cultural life of the Andes. In an ethnographic piece published in *La Prensa* (Buenos Aires) on 12 January 1941, Arguedas unambiguously described the effect of the highways on the traditional fairs of the highland: "the highways are strangling these fairs" ("Las carreteras están ahogando estas ferias"; *Páginas* 167). With the increased mobility of goods on the highways, the system of small towns like Incahuasi and Pampamarca that served as markets has decreased in importance. As in his stories and novels, Arguedas here includes a response from the landscape as seen through the eyes of the Indians: "The older Indians of Lake Parinacochas complain and say that the auki ('chief') Sarasara is resentful that his fair gets more and more rundown, year after year; that he is angry, and that some moonlit nights he wanders all the pampas next to his snowy crest, and often sinks large boulders to the bottom of the lake, and that the water of the lake weeps as well, and complains about the silence in which the formerly great fiesta is now celebrated" ("Los indios viejos de Parinacochas se quejan y dicen que el *auki* Sarasara está resentido porque su feria decae año tras año; que está molesto, y que algunas noches de luna camina por todas las pampas vecinas al nevado, y que muchas veces hondea grandes peñascos y los hunde en el fondo de la laguna, y que el agua de la laguna también llora y se queja por el silencio en que pasa ahora la gran fiesta de antes"; *Páginas* 166).

As might be expected, bridges perform a communicative function analogous to that of roads. Because of their relatively short length and fragility, however, bridges are more prone to the dialectic of the doorway: like a door, a bridge can be open or closed. It signals both the division between two territories, and the possibility of communication or contamination between them. In *Deep Rivers*, Arguedas uses the bridge over the Pachachaca as a *leitmotif.* Most of the novel takes place in the "Spanish" town of Abancay, where Ernesto is being educated in a Catholic boarding school. Crossing the bridge means leaving the "Spanish" domain and entering Indian country.

At the end of the novel, Ernesto crosses the bridge to avoid the plague raging in the town. The stone, three-arched bridge has been closed to maintain the quarantine, symbolizing the isolation of the Spanish bridgehead from its Indian hinterland. Ernesto must cross instead on the hanging bridge: "The Pachachaca roared in the darkness, in the depths of the immense gorge. The bushes trembled in the wind. . . . 'I'll cross [the gorge], go to Toraya, and from there to the cordillera. The plague won't catch me.' I ran across the city. On the hanging bridge at Auquibamba, I crossed over the river in the afternoon. If the colonos, with their curses and their songs, had annihilated the fever, perhaps from the height of the bridge I would see it float by, swept along by the current, in the shadow of the trees" (*Deep Rivers* 233; "El Pachachaca gemía en la oscuridad al fondo de la inmensa quebrada. Los arbustos temblaban con el viento. . . . —La atravieso, llego a Toraya, y de allí a la cordillera . . . ¡No me agarrará la peste! Corrí; crucé la ciudad. Por el puente colgante de Auquibamba pasaría el río, en la tarde. Si los colonos, con sus imprecaciones y sus cantos, habían aniquilado a la fiebre, quizá, desde lo alto del puente la vería pasar arrastrada por la corriente, a la sombra de los árboles"; *Los ríos profundos* 247). Running at right angles to each other, bridges and rivers perform opposing functions. Yet, in a complication of the image, the river is also a bridge—"Pachachaca" in Quechua means "bridge over the world." We have seen previously how spring run-offs can brings something of the Andes to the coast. The ambiguities of communication and closure are contained in Ernesto's first description of the Spanish bridge: "The bridge of the Pachachaca was built by the Spanish. . . . I didn't know whether I loved the bridge or the river more" (80; "El puente del Pachachaca fue construido por los españoles. . . . Yo no sabía si amaba más al puente o al río"; 69). Ernesto's locution does not simply place the Spanish bridge *over* the river Pachachaca. Instead, it is genitive (del), making the bridge belong to the river. Nevertheless, a few thoughts later, Ernesto formulates a choice for himself between the two elements of landscape. The choice functions on two symbolic levels at once, since bridge and river serve both as metonyms (for Spanish and Indian cultures) and metaphors (for the ambivalence of cultural communication).

In this invocation of cultural duality, the novel returns to another bridge/river: the Inca Wall in Cuzco. It is important to note that the Inca wall's magic for Ernesto derives not just from its representation of indigenous culture and the power and artistry of the Inca civilization, but also from its ability to support the newer, colonial architecture built upon it. The wall thus symbolizes the simultaneous existence of two cultures in a situation of interdependence. It is a bridge between these, as well as between past and present:

> When my father indicated the wall, I paused. It was dark, rough; its inclined surface was alluring. The white wall of the second story continued upward in a straight line from the Inca wall. . . . I walked along the wall, stone by stone. I stood back a few steps, contemplating it, and then came closer again. I touched the stone with my hands, following the line, which was as undulating and unpredictable as a river, where the blocks of stone were joined. . . . The stones of the Inca wall were larger and stranger than I had imagined; they seemed to be bubbling up beneath the whitewashed second story, which had no windows on the side facing the narrow street. . . . Couldn't one say *yawar rumi*, "bloody stone," or *puk'tik yawar rumi*, "boiling bloody stone"? The wall was stationary, but all its lines were seething and its surface was as changeable as that of the flooding summer rivers which have similar crests near the center, where the current flows the swiftest and is the most terrifying. (*Deep Rivers* 3, 6-7)

> Cuando mi padre señaló el muro, me detuve. Ero oscuro, áspero; atraía con su faz recostada. La pared blanco del segundo piso empezaba en línea recta sobre el muro. . . . Caminé frente al muro, piedra tras piedra. Me alejaba unos pasos, lo contemplaba y volvía a acercarme. Toqué las piedras con mis manos; seguí la línea ondulante, imprevisible, como la de los ríos, en que se juntan los bloques de roca. . . . Eran más grandes y extrañas de cuanto había imaginado las piedras del muro incaico; bullían bajo el segundo piso encalado que por el lado de la calle angosta, era ciego. . . . ¿Acaso no podría decirse "yawar rumi", piedra de sangre, o "puk'tik yawar rumi", piedra de sangre hirviente? Era estático el muro, pero hervía por todas sus líneas y la superficie era cambiante, como la de los ríos en el verano, que tienen una cima así hacia el centro del caudal, que es la zona temible, la más poderosa. (*Los Ríos Profundos* 8, 10, 11)

The process of dissolving the hard stone into flowing water imagined by Ernesto passes of necessity through the Quechua language. The equation of the wall with a "bloody boiling river" anticipates the novel's apocalyptic ending, where Indians are shot by civil guards and fall into the river. "Blood" also refers to the violence and suffering that constitute an integral part, as it were, of the wall's double construction, of Spanish dominance over Indian culture. To narrativize this feature of the landscape reveals that the motion Ernesto sees in the wall is an epiphenomenon of the "bloody" narratives it contains, of which the events of *Deep Rivers* will provide further examples, such as the native women's rebellion against the salt tax, and the plague that descends on Abancay as if in response. However, "yawar mayu" can also be translated as "river of blood," in which case it means something similar to English or Spanish "roots," the territorializing agency of a race or culture to which one belongs.

In his provocatively titled, *What Time is This Place*, geographer Kevin Lynch reproduces a photograph of the very Inca wall that Ernesto caresses, with the following caption: "Inca masonry lines the streets of Cuzco in Peru, still carrying the modern structures above, still proclaiming the past" (58). He chooses this as an example of a "time-deep" area: "The contrast of old and new, the accumulated concentration of the most significant elements of the various periods gone by, even if they are only fragmentary reminders of them, will in time produce a landscape whose depth no one period can equal" (57). This ability of landscape to give an indirect image of time, Lynch argues, explains our aesthetic responses to it. The uncomfortable feeling aroused in most people by housing developments, even when the structures and landscaping themselves are not ugly, comes from the opposite effect: since all structures and features were constructed simultaneously, they obliterate our notion of time. The Inca wall is to time what the bridge is to space. In each case, the object makes visible a culture considered by the dominant one to be past and invisible. Its opposite is the *barriada* ("shantytown"), for example, of the port city of Chimbote, where structures arise haphazardly and whose frontiers enlarge almost daily as it receives its flood of workers from the Andes. Arguedas was fascinated by the *barriada* and by its possibilities. He took numerous photographs of its structures and topography that appear at the head of the edition of *El zorro* in the *Obras completas*. One of the photos shows a reed house with a red flag waving from the roof. The flag signals that the house dispenses *chicha*, an alcoholic beverage of prehistoric origin, made from fermented corn. While we hear in the novel itself of Indians becoming accustomed to new forms of life such as whiskey and prostitution, Arguedas seems to have taken the photograph to demonstrate a more active and balanced process of transculturation. In the novel itself, however, transculturation is more frequently encountered in the form of prostitution, which reterritorializes sexuality as an exchange commodity.

The ambiguity of connecting structures in Arguedas contributes to the dynamism of his thought. Arguedas's fiction and anthropology, as Silvia Spitta has demonstrated, draws a mental map not dividing Indian from European, but placing into relation various kinds of "mestizo-in-process," social groups at different stages of becoming. Arguedas took care to insert the changing material conditions of communication between these two groups into his novelistic world. Nowhere is the "mestizo-in-process" more clearly presented than in his last novel, *El zorro*. It is shown, for example, in the ubiquity of the character Don Ángel, who takes on the mythic outlines of one of the foxes. The words he uses to describe himself, up to and including the relative neglect of the Amazon region, could apply just as well to his creator, Arguedas: "I am of the whole coast: dunes; rivers; villages, Lima. Now I am from above and from below, I understand the mountains and the coast, because I speak with a brother I've had for a long time in the highlands. I don't know anything about the jungle" ("Yo soy de toda la costa, arenales, ríos, pueblos, Lima. Ahora soy de arriba y abajo, entiendo de montañas y costa, porque hablo con un hermano que tengo desde antiguo en la sierra. De la selva no entiendo nada"; 77). Don Ángel becomes Peru itself, it would seem, by uniting its various regions.

The image of a river in this novel appears not in its crystalline, anthropomorphic form, but as a *lloqlla*. As we have seen, Arguedas used this image as early as "Orovilca," but without the Quechua name. As Don Ángel explains to the other fox, from above, the flood of humanity into Chimbote constitutes a *lloqlla*:

> Who the devil can put a *lloqlla* into a mold? Do you know what a *lloqlla* is? The avalanche of water, earth, tree roots, dead dogs, and rocks that descend churning beneath the current when the rivers are filled with the first rains in these stupid mountains. Chimbote is like a *lloqlla* now, listen: and no one knows anyone here. I told you we had reduced our workforce from two hundred fifty-eight down to ninety-six, right? This *lloqlla* eats hunger. The more workers we lay off, the more arrive from the mountains. And the shantytowns grow and grow, and marketplaces appear with more flies than food. Happy, happy, happy the seagulls with the death that surrounds them and with the avalanche *lloqlla* of life that surrounds them.

> ¿Quién, carajo, mete en un molde a una lloqlla? ¿Usted sabe lo que es una lloqlla? —La avalancha de agua, de tierra, raíces de árboles, perros muertos, de piedras que bajan bataneando debajo de la corriente cuando los ríos se cargan con las primeras lluvias en estas bestias montañas. . . . —Así es ahora Chimbote, oiga usted; y nadie nos conocemos. Le dije que redujimos los obreros de doscientos cincuentiocho a noventiséis, ¿no? Esta lloqlla come hambre. Más obreros largamos de la fábricas más llegan de la sierra. Y las barriadas crecen y crecen, y aparecen plazas de mercado en las barriadas con más moscas que comida. ¡Felices, felices, felices los alcatraces con la muerte que les ronda y la avalancha lloqlla con la vida que les ronda! (*El zorro* 77)

The metaphor of the *lloqlla* becomes more poignant if one recognizes its lack of fit with the extreme dryness of the coastal areas. Also, if one layers this image on top of the various mentionings of road-building in Arguedas, it becomes clear that road and river are now one thing. As in previous works, Arguedas shows a preoccupation with hydrology and with the extreme differences in climate and topography between the highlands and the coast. Here, however, his imagery has become more violent, and there is a recognition that the river mixes rather than separates. Indians, blacks, mestizos, Europeans, Americans, all are caught up in this whirlpool and beaten together by economic and demographic forces no one can control. The space of Chimbote differs tremendously from the regions marked and traced by rivers. It is a delta, the mouth of a river where the detritus is deposited. Arguedas deliberately incorporates into the novel, as Antonio Cornejo Polar points out, "the most diverse human and social destinies. They all flow together ['confluyen'] into a single space: the port of Chimbote, reality and symbol" (278). The coast is no longer a place to commute to for work and then return to the highlands or die by the side of the road back, as depicted in *Yawar fiesta*, but a site for the reterritorialization of Indian identity freed from its feudal constraints and supports. One would expect this process to result in the usual feared outcomes of globalization: the loss of traditional folkways in favor of standardized customs. In his anthropological work, Arguedas himself expressed these fears vis-à-vis music and artisanal production. Yet, as Cornejo Polar demonstrates (269-78), the title of the novel, *The Fox From Above and the Fox From Below*, attempts to work a dialectic by encapsulating the story of Peru's globalization within a Quechua myth. On the other hand, the possibility provided by the narrative—that Native tradition may prove more durable and transformative than the processes of capitalism that seem to be destroying it—contrasts with the interpolated diary entries of the novel that reveal Arguedas's weariness and despair. This oscillation belongs, as it were, to the never-ending process of reterritorialization in Peru, which is, in Nikos Papastergiardis's words, "never complete or final, but a constant oscillation between the other states. In the exilic search for home, [Deleuze and Guattari] cannot even determine which comes first. For the very act of seeking a territory implies a prior deterritorialization and its own reterritorialization in the realm of myth and dream, or land and commodity, with each movement occurring not in a linear sequence but within a knotted entanglement" (118). That is, within a *lloqlla*. Thus, the *lloqlla* or river of mud replaces the *yawar mayu*, or river of blood, that had symbolized rootedness in Arguedas's earlier work. It is difficult to determine whether this change of metaphor reflects an overcoming of an excessively polarized and static vision of reality, or a succumbing to the forces of history. It is equally difficult to distinguish between Arguedas's personal development over the course of three decades as an author and visionary of Peruvian nationalism, and the experienced acceleration of those historical changes over the same period. Increased migration, industrialization, urbanization, and globalization may have become so pronounced during these decades that Arguedas could no longer ignore them. Whatever the cause, his work shows throughout its long trajectory an attempt to understand and communicate these processes through the features of the Peruvian landscape.

Postscript

29 July 2001: Alejandro Toledo, the first Peruvian president from an indigenous background, traveled to Macchu Picchu to receive the blessings of the *waki*, the same mountain gods so prevalent in Arguedas's writing. In Cuzco, thousands of Indians in traditional dress watched the ceremonies on wide-screen television. Toledo deliberately invoked the mixing of old and new, traditional and technological, local and globalized in contemporary Peruvian society. What would Arguedas have thought of this moment in Peruvian history? Would he have seen it as a healing moment, a triumphant reversal of the centuries of blindness of one part of Peruvian towards the other half so painfully documented in his work? Or would this master of literature have seen the great rift between the symbolic level of state institutions and the inexorable flood of globalization carrying most of Peru's population out to sea?

Chapter Seven

Voces Clamantes and the Desert Landscapes of Austin, Brossard, and Harjo

In the last three chapters, I have argued that the tortuous narratives of José Maria Arguedas, Juan Benet, Euclides da Cunha, and William Faulkner arise out of these authors' problematic, divided, or contradictory conceptions of nation, and its entanglements with regionalism. The other authors considered so far have almost insisted on the stability of their true imaginary places, although one might question their readers' ability to move from these authors' mimeses of place to their own lived reality within the referred nation or region. This last chapter compares three works that deconstruct such notions of landscape's stability. Not surprisingly, they substitute notions of national identity with questionings of the notions of "in" groups and "out" groups, and then trouble these even further with issues of gender. Mary Austin's *Land of Little Rain*, Joy Harjo's *Secrets from the Center of the World* (with Stephen Strom), and Nicole Brossard's *Le Désert mauve* ("*Mauve Desert*") are all set in the American Southwest. Each author holds a different relationship to her chosen desert's landscape. Austin was originally from the Midwest, and nearly starved in California's Owens Valley before learning to find its hidden treasures. Harjo, a member of the Muskogee Nation, grew up in the arid environments of Oklahoma and New Mexico. Nicole Brossard, on the other hand, is a Québécoise writer. Her first experience of the Arizona-Sonoran desert was the shocked view of someone wondering where all the trees and buildings and clouds had gone, leaving language to run at full velocity.

I show how the desert landscapes of these texts bear a relationship to the gender of their authors and that both these variables in turn produce a deconstruction of nation. Kate McCullough has observed that nineteenth-century US women writers were the first regionalists to go beyond New England, confronting with their close observation of Mexican Americans or Cajuns the myth of a homogeneous, white America. Similarly, "the twentieth-century literary critical dismissal of women's 'regional fiction' . . . is one manifestation of an American literary enactment of nation building, a means of suggesting that 'we' have a single and unified literary tradition" (4). A growing body of feminist scholarship argues that "nationalism is constituted, from its origins, as a highly gendered relationship, in which women are not 'imagined' to be national citizens" (Radcliffe and Westwood 135; see also McClintock; Kandiyoti). It is no accident that deserts also lay outside the traditional notion of landscape as a source of aesthetic experience for centuries. Naturally, the generally inhospitable nature of the desert leads to low population density. However, it is worth noting that a similarly low population density in the Amazon rainforest has only piqued the world's interest in that area. Mountains are similarly uninhabitable, but they became an object of aesthetic curiosity long before deserts did. Jay Appleton's prospect-refuge theory of our enjoyment of landscape provides an explanation of this inadequacy of desert. Whereas "habitat theory postulates that aesthetic pleasure in landscape derives from the observer experiencing an environment favourable to the satisfaction of his biological needs . . . prospect-refuge theory postulates that, because the ability to see without being seen is an intermediate step in the satisfaction of many of those needs, the capacity of an environment to ensure the achievement of this becomes a more immediate source of aesthetic satisfaction" (66). In other words, human's general preference for forested or vegetated environments goes beyond the pragmatic aspect of survival to the same instinct for refuge that drives your cat to seek hidey-holes. You can hide in jungle and forest, and from the top of a mountain you can look down on your pursuers. In the desert, these advantages are lost. You gain at best a draw, as your gaze meets that of the Other across a distance.

Historically, those defying the rule of desert avoidance have been saints, holy people, or visionaries. These are, of course, the exceptions that prove the rule, for their sojourn in the desert functions to deprive them of human culture and to bring them closer to God. Yi-Fu Tuan has attempted to formulate the "ambivalent aesthetics" of the desert as a philosophy of asociality and tie it to a national ideal: "In North America and Australia, the sweeping spaces of the dry interior have been elevated to mythic status. Frontier and outback have become national symbols of hardy responsible manhood, of individualism in America and camaraderie in Australia, of a clean and genuine way of life inspired by nature and the spirit of place, in contrast to the parochialism and communal stickiness, the unassimilated alien ways of the coastal cities. In both countries the love of the interior was fueled by misogyny, a distaste for the softness of culture, for commercialism, and more generally mankind in its swarming numbers" ("Desert and Ice" 145). Misogynistic? Perhaps. What if, on the other hand, the desert were to constitute a "line of flight" from patriarchal culture? If the absence of people also meant an absence of structure and the challenge to reformulate language? The image of "howling wilderness" is most characteristic of Puritan thought, and changed in substantial ways through the nineteenth century, as documented by Robert E. Abrams and David Teague. In fact, as Teague has argued, the mental map of the American Southwest was reshaped at the turn of the twentieth century: "Until about the mid-1890s, deserts were powerful icons of howling wilderness for citizens of the United States. They were incomprehensible to the collective imagination of the 'civilized' portion of the country because they were places where accustomed modes of geography, agriculture, industry, and commerce did not obtain. But by 1910 deserts had become associated with the very height of American culture" (3). Audrey Goodman links the transformation of desert affect in the American psyche to increased urbanization and industrialization: "the forces of capitalist expansion . . . lowered spatial barriers only to create a demand for diverse places; the rise of corporate, scientific, and artistic specialization as well as a psychology of individuation, reinvention, and escape" (xiii).

Goodman's three terms—individuation, reinvention, and escape—all point toward a rejection of historical continuity, a feature of the desert landscape that has fascinated Jean Baudrillard as well. In his meditation on US culture, *Amérique*, Baudrillard argues that American landscape remains at the level of simulacrum rather than symbol. Baudrillard literalizes the metaphor of utopia. In the oxymoronically entitled chapter "Utopia Achieved," Baudrillard critiques a French philosophy that always saw utopia in the idea of America, but never in its actual practices. Baudrillard's narrative "use[s] manifest refutations of environmental determinism, or a discontinuous geography, to illustrate the denial of historical determinism, or the historical discontinuity, that lies . . . at the heart of American modernism" (83). The utopia achieved, Baudrillard insists, is local, practical, a form of *bricolage* without anterior concepts. Desert is the landscape of America, because in the desert things exist for themselves, critiquing human signifying practice. Austin, Harjo, and Brossard all write texts that participate in such a critique. In these texts, the mapping process is taken to a level of self-reflexivity: what is mapped is above all the different possibilities of orientation in a smooth space. Another way of expressing this difference, following Goodman's suggestion, is to compare the process of diataxis with that of translation, as these texts move from mimesis to diegesis.

Landscape lines

Throughout her forty-year writing career as poet, dramatist, essayist, and folk arts specialist, Mary Austin sought a theory of culture that would link literature to the land we live on. Born in Illinois in 1888, at the age of twenty Austin moved with her family to central California. She suffered malnutrition on the family farm, until she discovered wild grapes that became for her a gateway into her natural surroundings that fascinated her from that moment on. David Wyatt draws an interesting parallel between Austin's apprenticeship in supplying her own nutritional needs from the landscape's hidden treasures and her realization that the very sparseness of that landscape would feed her imagination, and that the resulting narratives would be minimalist: "The actual 'malnutrition' of the body Austin suffered during her first months in California had looked like a fate also in store for her mind. Her discovery that she could recover physical health with 'something grubbed out of the woods' (wild grapes) corresponded with her discovery that it was the very poverty of the land on which she could feast her spirit. The imagination's strength came to be measured by its power to nourish the half-starved" (93). The region's arid landscapes formed the basis for Austin's best writing. Indeed, her very first collection, *Land of Little Rain*

(1903), today perhaps more popular with environmentalists than with literati, is her only volume to have remained continuously in print.

Austin's supposed environmentalism was really just an outgrowth of her conviction as to the inescapability of landscape's influence on culture. She believed "that America really had no choice about adaptation to the land, that in the long run through the process of mutual adaptation of the land and the people, the land always won" (Rupert 377). I have noted a similar ideology at work in Turgenev, Da Cunha, and Arguedas. Austin's inspiration was the Native American cultures she studied, for whom adaptation to the natural environment was a given. By 1915, Austin had sketched out a theory that dance drama constituted the *Gesamtkunstwerk* of native peoples, and that she could read the origins of any song-dance, without knowing the language it was sung in, by studying the piece's rhythm and identifying the landscape that had given it shape. This particular form of diataxis she called the "landscape line," a "leap of the running stream of poetic inspiration from level to level, whose course cannot be determined by anything except the nature of the ground traversed" (*American Rhythm* 55). Austin did not define the term precisely, nor did she provide enough examples to clarify it completely, but James Rupert has helped us understand the concept as "a function of how the thought of the line reflects the landscape in form" (385). Austin's most nuanced example describes the Hako ceremony of the Pawnees. Each stage of the journey described in this ceremony receives a different rhythm, according to the landscape crossed: "As the visiting party draws up from the lowlands about the river, we have this finely descriptive rhythm: The mesa see, it's flat top like a straight line cuts across the sky. It blocks our path, and we must climb, the mesa climb. What work in any language more obviously illustrates the influence of environment on literary form? Other examples there are of much subtler and more discriminating rhythms, but they only announce themselves after long intimacy with the land in which they develop. The homogeneity of the Amerind race makes it possible to detect environmental influences with a precision not possible among the mixed races of Europe" ("Non-English Writing" 630).

Austin's concept of the landscape line finds confirmation in our understanding of the "songlines" used by the Aborigines of Australia to "map" their land. They sing songs that tell the stories of ancestors who, in their own journeys in the Dreamtime, created the natural features of the landscape. According to Bruce Chatwin, "In theory, at least, the whole of Australia could be read as a musical score. There was hardly a rock or creek in the country that could not or had not been sung. One should perhaps visualise the Songlines as a spaghetti of Iliads and Odysseys, writhing this way and that, in which every 'episode' was readable in terms of geology" (13). Or, as Austin puts it, every landscape is readable in terms of its singing. The songlines also constitute, in large part, the "literature" of the Aborigines. As did Da Cunha in Brazil and Arguedas in Peru, Austin recognized that the view of what an American national literature might look like is hopelessly distorted by a neo-European viewpoint that applies criteria created in another continent to the Americas. Austin argues that the construction of a truly "national" American literature must begin with the placement of aboriginal creation such as the Hako at the center of the canon, according to the robustness of its landscape line. She concludes her article on "Non-English Writings" with the following remarkable paragraph: "It is not . . . the significance of Amerind literature to the social life of the people which interests us. . . . The permanent worth of song and epic, folk-tale and drama, aside from its intrinsic literary quality, is its revelation of the power of the American landscape to influence form, and the expressiveness of democratic living in native measures. . . . in every country in the world . . . the really great literature is found to have developed on some deep rooted aboriginal stock. The earlier, then, we leave off thinking of our own aboriginal literary sources as the product of an alien and conquered people, and begin to think of them as the inevitable outgrowth of the American environment, the more readily shall we come into full use of it" (633-34).

When creative writers turn critics of literature, the theories they develop tend to confirm or explain their own writing practice. Thus, we might expect to find some landscape lines in the *Land of Little Rain*. Austin herself found them in her desert writing, when she reread it: "when I had occasion to turn back to one of my desert books, written eighteen years ago and scarcely looked at since, I discovered the first paragraph striking without intention into the irregular tug and release of the four horse Mojave stage and of the eighteen-mule borax team" (*American Rhythm* 14-15). Another line can be discovered by tracing the routes taken by Austin out of her home base of Lone Pine to the scenes of various stories. The slow, microclimatic movements of these into the hills and valleys to the east and west contrast with the speed and power of the main stagecoach moving north and south, up and down the Owens Valley. Trails and borders are the constant metaphors that organize Austin's writing. For example, Austin offers each of the stories in *Lost Borders* as a trail into the desert, radiating out from Lone Pine (see Figure 19):

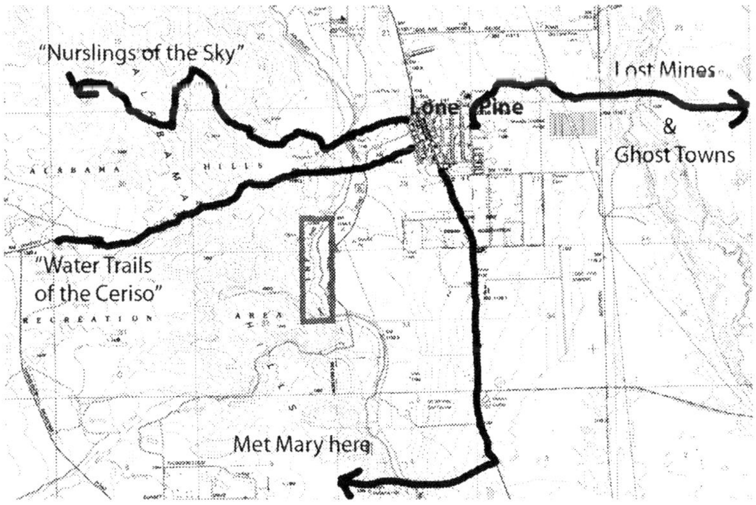

Figure 19. 1883 Topographic map of the Lone Pine quadrangle, with Mary Austin's travels drawn in.

"All the trails in this book begin at Lone Pine, winding east by south and east again, although you will look long without finding the places where things happened in them unless you are susceptible to those influences that contribute to the fixed belief of mining countries, that the hot essences of greed and hate and lust are absorbed, as it were, by the means that provoke them, and inhere in houses, lands, or stones to work mischief to the possessor" (*Stories* 161). The idea is Faulknerian (or Faulkner, the later writer, is Austinian) in several ways. The first is the microclime. The broad, monotonous vistas of the desert deceive the viewer, Austin implies; they blind us to the small pockets that may unfold into diegesis. This, we recall, was also Gogol's mimesis in *Dead Souls*. Secondly, in Austin as in Faulkner, our mental maps must be color-coded with greed, hate, and lust for us to find our way in a landscape that may seem empty.

The initial sentences of *Land of Little Rain*, deliberately rhythmic, also proclaim a theory. They argue Austin's preference for the Indian way of name-giving, in which the name is relative to the name-giver's point-of view: "every man [is] known by that phrase which best expresses him to whoso names him. Thus he may be Mighty-Hunter, or Man-Afraid-of-a-Bear, according as he is called by friend or enemy, and Scar-Face to those who knew him by the eye's grasp only" (xvii). Indian names are not meant to fix, differentiate, or perpetuate, but rather to mediate between object, namer, listener, and circumstance. They are pure diataxis, translations of experience into hypothesis. Austin immediately adopts this practice in her own writing, to define the Owens Valley and its environs. The title of the first sketch and of the collection is "Land of Little Rain." However, the first sentence gives this region another name: "East away from the Sierras, south from Panamint and Amargosa, east and south many an uncounted mile, is the Country of Lost Borders. . . . Not the law, but the land sets the limit. Desert is the name it wears upon the maps, but the Indian's is the better word" (1). As the boundaries of this region have been lost, so too its linguistic references, which slip between various alternatives. The perspectivism inherent to this way of naming shapes Austin's stories, where the teller invariably receives more attention than does the tale. This situatedness shapes the meaning of her stories. By invoking this possibility of a nonmathematical, nonchorographic mapping through a polyphony of names, Austin raises the questions of where her own book belongs, and of whether the reader will be friend or enemy to these stories that serve as names for the places of the land-of-little-rain. Indeed, Austin's renaming of California's Owens Valley with the epithet of her title reflects this perspectivism and the ineffability of the desert landscape: "there are certain peaks, cañons, and clear meadow spaces which are above all compassing of words, and have a certain fame as of the nobly great to whom we give no familiar names" (*Land* xvii). This aspect of translation has been noticed by Austin critics, one of whom describes the landscape of little rain as a text that possesses translatability: "Through [Austin's] imaginative revision the land is exposed as the site of ineffable mysteries—a locus that 'breeds' language forms equal to the task of translating it for ordinary understanding" (Stineman 71). A name is also a trail; just as several trails converge on a point, so too the several names for a thing converge on it. Trails form a constant theme throughout Austin's work. Usually, these are trails that are lost, dead-ended, or too faint to see with the naked eye, like the stories the desert tells. In Faulkner's hill country, the past continues into and strangles the present, as in "A Rose for Emily," about a woman who keeps her dead lover's corpse in her house for decades. A narrative form of this is the ghost story, a type of tale Faulkner loved to tell. The basic narrative of the desert, on the other hand, is the "lost mine" story, the incomprehensible loss of the past and recognition of the discontinuity of time. Ghost towns yield few ghost stories: as Lois Rudnick reminds us, Austin depicts consistently women as being more at home in mapless areas than men: in the "*Land of Lost Borders* (1909), where her 'outliers' live, women are liberated from the restrictive norms of femininity by the vast and undominated scale of the land" (18). Furthermore, throughout the *Land of Little Rain*, Austin presents herself as the mapmaker of trails too faint for men to see. The "Water Trails of the Ceriso" are a geo-glyph engraved in the land, "faint to mansight, [but] sufficiently plain to the furred and feathered folk who travel them" (9). Significantly, "Man-height is the least fortunate of all heights from which to study trails" (9). The complex gendering of desert activity and interpretation runs throughout Austin's work. In a story from *Lost Borders*, she writes of a minister who encourages a woman to stay with her husband: "The minister himself was newly from the East, and did not understand that the desert is to be dealt with as a woman and a wanton; he was thinking of it as a place on the map" (186). Man-height and the chorographic, European map cause blindness. To the woman, Austin, however, closer to the ground and to the sky than a man, the trails reveal themselves.

The trails are mirrored almost exactly by the hieroglyphs of the heavens, which Austin describes with a maternal metaphor as "Nurslings of the Sky": "Weather does not happen. It is the visible manifestation of the Spirit moving itself in the void" (90). Men seem to have missed not so much the signifiers this time, as their relation to this ultimate signified of spirit. Instead, men link each weather form to an earthly occupation which may benefit themselves: "You will find the proper names of these things in the reports of the Weather Bureau—cirrus, cumulus, and the like—and charts that will teach by study when to sow and take up crops. It is astonishing the trouble men will be at to find out when to plant potatoes, and gloze over the eternal meaning of the skies" (96). Weather charts are maps meant to navigate time rather than space, and in noting their inadequacy, Austin also rejects every form of chorographic reduction.

That land, flora, and fauna could ever be more than just instruments in the making and carrying on of life needs explaining, and Austin broods on that explanation. Here is one such meditation, from "Other Water Borders": "I came first upon a wet meadow of yerba mansa, not knowing its name or use. It looked potent; the cool, shiny leaves, the succulent, pink stems and fruity bloom. A little touch, a hint, a word, and I should have known what use to put them to. So I felt, unwilling to leave it until we had come to an understanding. So a musician might have felt in the presence of an instrument known to be within his province, but beyond his power. It was with the relieved sense of having shaped a long surmise that I watched the Señora Romero make a poultice of it for my burned hand" (*Land* 85). We move through a variety of semiotic systems in this mini-narrative: the appearance, taste, and smell of the plant; the tactile apprehension of the poultice made from its leaves; language itself, the translated word that Austin awaits from the plant; and the sound of music in a simile for Austin's own writing process. Austin could have written most of the chapters of *Land of Little Rain* as first-person memoirs, for they all recount her own encounters with the land and its inhabitants. Similarly, Austin constantly reminds us that storytelling is not her purpose. In the chapter "Jimville," about a mining town, a ghost-town-in-process we might say, she negatively evokes the stories of Western writer Bret Harte. Jimville would be a town for him to draw diegesis from the mimesis of its ruins; the inhabitants are compared with his characters, but Austin sees her task as a different one. "If it had been in medieval times you would have had a legend or a ballad. Bret Harte would have given you a tale. You see in me a mere recorder, for I know what is best for you; you shall blow out this bubble from your own breath" (42). Austin prefers to furnish the reader with a mimesis and this simple command to diataxis that equates it with a living breath. There is irony here, for Jimville is as fictional as any of Harte's towns. It is a composite drawn from a range of geographical data and a wealth of stories. Austin's mimesis is itself a bubble. Unlike Harte's locations that are scenes for narratives to happen in, Jimville grows out of its desert environment, and this growth forms the whole of the narrative. A recorder would be merely of the language of the inhabitants as language, or, if a photographer like Ansel Adams, who illustrated a 1950 edition of *Land of Little Rain* with his photographs, one who captures images.

"Jimville" is just one of many Austin sketches that is more interested in contemplating its own narrative impedance—thus foregrounding its diataxis—than in completing a tale. If plotline is a river, then Austin's tales are springs and oases—the half-told story of the "Walking Woman" bubbles out of her beside the waters of Warm Spring. According to Austin's own theory, we must look to the desert environment for an explanation of her reluctance to narrate. In fact, she herself provided the analogy in an early story, "The Lost Mine of Fisherman's Peak":

> This is a large country, with few and far-between oases of richness and greenness. One may take days' journeys in it and not come by any place or occasion whereby men might live; and other days stumble upon the wealth of dreams. Weeks on end the traveler finds no towns nor places where towns could be, and then drops suddenly into close hives of men, digging, jostling, fighting, drinking, lusting and rejoicing. Every story of that country is colored by the fashion of the life there; breaking up in swift, passionate intervals between long, dun stretches, like the land that out of hot sinks of desolation heaves up great bulks of granite ranges with opal shadows playing in their shining, snow-piled curves. (*Western Trails* 210)

In particular, the last sentence of this passage is yet another formulation of the theory of landscape line. The mental map that Austin draws in this passage meanders and envelops nothing. Identity, natural or otherwise, is "on the line." Austin's essays and stories parallel the weeks on end in the desert, while the incidents and point of her stories are digging, jostling, and rejoicing.

The center of the line

Joy Foster was born in 1951 in Tulsa, Oklahoma, but attended high school and college in Santa Fe, New Mexico. In 1970 she took the last name of her paternal grandmother, a Muskogee. Joy Harjo's and Stephen Strom's *Secrets from the Center of the World* (1989) consists of a brief introduction and a series of prose poems by Harjo describing photographs of the Arizona landscape taken by Strom. The book functions both as literature and as theory; it simultaneously provides a picture of the landscape, and a heuristic for "viewing" that landscape. It deliberately resists the landscape practices of European and American painting, and later photography. Harjo attempts neither to reproduce nor to supplement the iconicity of the photograph; rather, her prose moves contrary to and tries to escape Strom's visual vectors and square frames. The landscapes, for example, are entirely devoid of people, and so Harjo "blows out the bubble" of diegesis with her own breath, translating into English the stories hidden in the land. In creating tension between picture and logos, this brief book becomes an Austinian exploration of the landscape line. It shows that "the earth . . . speaks stories. They are of its own origins as the keeper of bones; of survival; of the travels and changes of the people moving on it, inside it; of skies. And the stories change with light, with what is spoken, with what is lived" (*Secrets* n.p.). What happens in this text, then, is a complicated, three-way translation between earth, photograph, and word.

An example is the following text and its accompanying photograph (see Figure 20)—or is it vice versa?: "Stories are our wealth. Winter nights we tell them over and over. Once a star fell from the sky, but it wasn't just any star, just as this isn't just any ordinary place. That cedar tree marks the event and the land remembers the flash of its death flight. To describe anything in winter whether it occurs in the past or the future requires a denser language, one thick with the promise of new lambs, heavy with the weight of corn milk" (24). In an era of DNA and of the creation of materials that "remember" their shape and form, it should come as no surprise to learn that the land, the earth, also remembers and narrates. The first two sentences of the story about stories are meta-discursive, inviting and challenging the readers to identify themselves with the "we" in the practice of storytelling and story-listening. Like the games Benet and Faulkner play with the simultaneous identification of and disjunction between region and nation, so too Harjo makes the reader wonder whether he is also a listener. Most of us fail the test, cannot identify with this particular cultural practice. Walter Benjamin's famous distinction between "story" and "information"—"it is indispensable for information to sound plausible. Because of this is proves incompatible with the spirit of storytelling" ("The Storyteller" 89)—seems to separate us irrevocably from this process. In Baudrillard's terms, "our" winter nights are filled with television sets that are constantly on, although no one is watching.

Figure 20. Harjo/Strom's graphic poem from *Secrets from the Center of the World* 24.

The stream of language, sound, and images from the media is for a form of silence, leading neither to a "global village" nor to an "imagined community." Telling stories must be of a different cast, and so most readers end up interpreting the "we" of these first sentences as Indian, as aboriginal, also in Austin's sense of a literature that arises from the land.

A second example is somewhat less meta-discursive: "Two sisters meet on horseback. They gossip: a cousin eloped with someone's husband, twins were born to his wife. One is headed toward Tsaile, and the other to Round Rock. Their horses are rose sand, with manes of ashy rock" (42). Here, both photographer and story-teller engage in minimalism. In fact, if the photograph (see Figure 21) is regarded outside of the context of the book—apart from its translation—then it is difficult to determine its top and bottom (I became dizzy looking at the black-and-white glossy I had made of it for this study). A careful look at the photograph, however, reveals a whirlpool shape in its center, that may also be thought of as two heads facing each other, the two sisters. But would a viewer be capable of finding this "hidden picture" without Harjo's accompanying narrative? We are back at Austin's idea of multiple naming; the minimalist narrative throws out shadows that suggest alternatives, like a good translation. Harjo repeats a process that is coterminous with the invention of language itself. Landscape and meaning come into being simultaneously, Siamese twins embodied in myth, an act of translation. Harjo's story, which seems to emerge from the earth, exemplifies the structure of this text-picture. The photography that captures landscape from the outside, as it were, combines uneasily with a *logos* that enables us to unite the accidental contiguities in the midst of which we find ourselves. Such is the process of narrativization by which most of us find the center of the earth, not by the mathesis we expect and trust. In Harjo's landscape there are no accidents. It should also be noted that the audience for this translation project, like most audiences for translation, is unfamiliar with the original culture, which is the landscape.

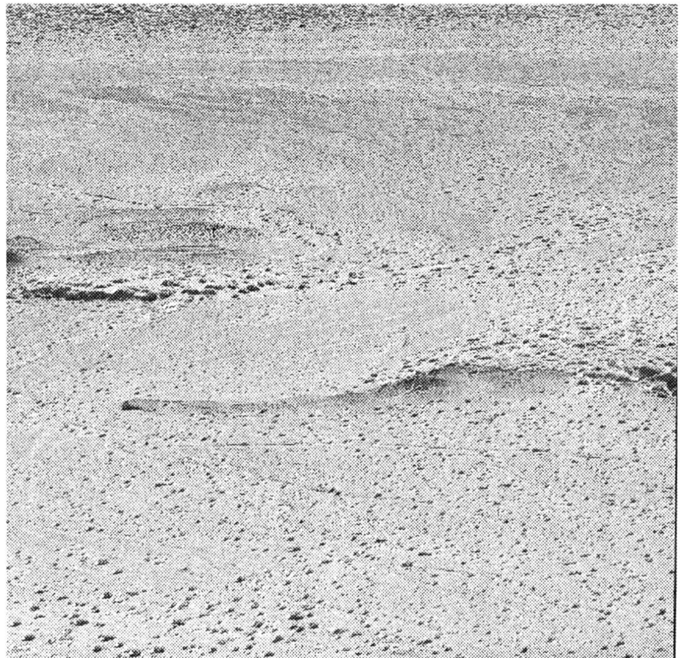

Figure 21. Harjo/Strom's graphic poem from *Secrets from the Center of the World* 42

Lines of flight

Nicole Brossard was born in Montréal in 1943. While she has spent most of her life in the province of Québec, Brossard explicitly opposes regionalist writing, in favor of an exploration of language that can change reality rather than merely reproduce it. Her French-language writings have been translated into English, German, Spanish, and Italian, and have gradually become part of the canon of lesbian feminist literature. One could speculate from reading Brossard's novel, *Le Désert mauve*, that the Québécoise author had read Austin and vowed to fictionalize the translation process represented in the *Land of Little Rain*. Indeed, Mary Austin's loneliness, which drove her to interact with the Mexican and Indian populations of the Owens Valley and to take long walks by herself, is modernized in the character of the teenager Mélanie, who drives through the Arizona desert, mostly at night. The French-French translation of *Mauve Desert* foregrounds the fusion between lesbian and heterosexuality, Québécois and Arizona landscapes, and adolescent and middle-aged mentalities. The book is divided into three parts, the first being Laure Angestelle's novel, *Le Désert mauve*, and the last Maude Laures's translation of the same, *Mauve l'horizon*. The middle section is a translator's notebook, where Laure records her notations on characters, settings, concepts, and her interview with the author. The book to be translated is a partial autobiography by Kathy's daughter, Mélanie. It concerns Kathy Kerouac and her lover, Lorna Mynher. Mélanie's passion for driving through the desert finally cathects on the "geometer" Angela Parkins, who is murdered by *l'homme long* (Longman), the only male figure in the work. For our purpose, we may think of Parkins's geometry as the attempt to find the Harjovian center of the earth and its secrets, destroyed by Longman's equations. We are meant to think of these as belonging to nuclear science, thus associating the desert with atomic bomb testing, the ultimate confirmation of the landscape's abjection. Nuclear weaponry, in turn, throws the shadow of national aggression over the desert's space of exile.

The concept of translation recurs in Brossard's work from the 1985 *L'Aviva*, in which Brossard provides French to French translations for ten poems, to her 1989 *Picture Theory*. Brossard defines the process of translation as "a complex act of passage which inflame[s] the mind." She claims that we translate "our emotions, our sensations, our values, that the words of others could only make sense if we translated them through our own experience, our referents, that part of life escaped us because we did not have the words to name, to translate, that is to embed into our experience the sensations, the feelings, the concepts which seemed so obvious in another language" ("Nicole Brossard" 53). An important referent for the pronoun "our" in this quote is women and their problematic relation to "man-made" language, symbolized in the text by Longman's equations. Brossard chooses the desert in a continuation of the paradox of ecstasy, which now is erotic rather than religious. It is now lesbian sexuality that finds luxurious exile in the wasteland: "The project to create a space for lesbian desire/subjectivity carries the author into realms of the unthought, cyprine islands suspended above the Ari(d)zona desert, beyond the textual/epistemological concerns of modernity" (Parker 108). The emptiness or smoothness of desert space means that any point provides numberless paths, or lines of flight, in any direction, similar to Brossard's idea of literature, in which readers do not recuperate an authorial intention, but rather use textual signifiers to create meaning. The desert's smoothness, conveyed throughout the Harjo/Strom book, would appear in Appleton's schema as simply another example of that biome's inability to furnish refuge or prospect. The desert is thus an *Angststelle* (German for "place of fear"), from which term Brossard derives the name of her fictional author who attempts its mimesis.

Like a detective, and unlike a geometer, the translator Maude Laures works on the principle of identification. Although both geometric reduction and translation are forms of mapping founded on notions of completeness, accuracy, and the ability to construct boundaries, the former works on distance and the separation of subject and object, the latter on their complete identification. This identification is symbolized in the overlapping of names—Laure and Laures—between author and translator. As the detective begins with the crime and works backwards to reconstruct its genesis, so the translator can begin with the fait accompli of the book, and attempt to map its "trajectory"—meaning simultaneously the jet of the story and the path of Mélanie's Meteor (the car she drives, symbolizing both the violence of atomic explosion and the ultimate line of flight, space travel) through the desert. Detective and translator alike are contaminated by this principle of identification, as it draws them closer and closer to the thoughts and language of the originator of crime and text: "Trajectory, thought Maude Laures, trajectory. And she progressively got accustomed to the idea of becoming a voice both other and alike in the world derived from Laure Angestelle. The characters would soon slip away one after the other, become little transparencies in the distance, crystallize. She would be alone in her language. Then would come substitution" (*Mauve Desert* 160; "Trajectoire, pensait Maude Laures, trajectoire. Et elle se faisait de plus en plus à l'idée de devenir une voix autre et ressemblante dans l'univers dérivé de Laure Angestelle. Les personnages allaient bientôt se défiler les uns après les autres, devenir de petites transparences au loin, se cristalliser. Elle serait seule dans sa langue. Alors, il y aurait substitution" ; *Le Désert mauve* 176). Maude Laures's hopes lie between hermeneutics and cartography. She attempts to reach the Archimedean point, a verbal "establishing shot" like Gogol's rider atop the Carpathians or the opening paragraph of *Os Sertões,* where the multivalence and polysemy of characters and events fade to become names on a map, reversing Austin's and Harjo's attempts at letting each microclime dig and jostle its own mimesis.

Laures visits the places of *Désert Mauve* and, like Austin and Harjo, embodies her experiences in prose poems on ideas such as "desert," "dawn," and "light." These, however, turn out to be *terrae incognitae*, holes in the map. One section, entitled "Laure Angestelle," imagines the author as several people. Angestelle's imagined age is progressively reduced from fifty to thirteen, and we are unsure whether she grew up in the Western desert (like Harjo) or moved there from the East (like Austin). The possible circumstances and reasons for the publication of this author's lone work are also outlined. But "all of this which could be fantasy in no way invalidated the thought that Laure Angestelle had no doubt been a proud woman with a supple body, eyes filled with torment, vulnerable in the face of beauty and silence" (83; "tout cela qui pouvait être fantaisie n'invalidait pas la pensée que Laure Angestelle ait sans doute été une femme fière, au corps agile, aux yeux pleins de tourment, vulnérable devant la beauté et le silence"; 90). Silence is the unchanging constant in Laure Angestelle, seen from all perspectives: "lest thinking start distinguishing between words, laughter, discourse, the culture medium, she needed silence, to put silence in front of beings like a screen for she knew the price of beauty to be the silence attuning all spheres of sound" (82; "avant que la pensée ne s'exerce à distinguer entre les paroles, les rires, le discours, le bouillon de culture, il lui fallait du silence, mettre du silence devant les êtres comme un écran car elle savait que la beauté était au prix du silence qui accordait toutes les musiques"; 88). The torrent of language, like the steady stream of sounds and images from Kathy Kerouac's television set, isolates rather than unites, creates rather than removes ambiguity. The silence invoked here is not the absence of sound. It is the silence of nature, of animals, of the desert, the very silence of impenetrability that makes Harjo and Austin so eloquent. Silence is the acoustic image of the desert's smoothness, and of the absence of refuge noted by Appleton. Author and translator achieve identity in this silence. Both "like dealing with silence but each one here is looking to understand how death transits between fiction and reality" (131; "toutes deux aiment composer avec le silence mais chacune ici cherche à comprendre comment la mort transite entre la fiction et la réalité"; 140). Laures is caught between her great desire to translate a text that fascinates her, and her inability to identify with the original's lesbian sexuality, not to mention with the kinds of description a fifteen-year-old would make of that sexuality. As if to allegorize this translation difficulty, Laures simply leaves Mélanie's pronouncement, "The desert is civilization" (18; "Le désert est la civilisation"; 18), out of her translation. This inability is equivalent to the incongruity of someone raised in the cold, humid environment of Quebec describing the hot, dry desert of the American Southwest. Like her lesbianism, Mélanie's pronouncement negates civilization, not only in and of itself—we need only reverse the position of the two nouns for that—but also in civilization's ability to pronounce the desert uncivilized and dead.

The vocabulary necessary for conveying the life of the desert, that which Mary Austin took years to acquire, and which Maude hopes she can fill her notebooks with, will be lacking, as it often is between languages. Brossard represents landscape above all as the site of bodily experience that influences thought. The contradictory and complex intervention of this landscape in the lives of the novel's characters escapes summary. Like Austin and Baudrillard, Brossard seems to have found in the desert a questioning of all previous reference points of cultural meaning. In the desert, things stand for nothing but themselves, posing the question: "how far can we go in the extermination of meaning, how far can we go in the non-referential desert form without cracking up?" (Baudrillard 10). Longman cracks up, as does the geometer Angela Parkins, as does the translator Laure Angestelle, as does the reader of Angestelle's translation, who will seek for the similarities between original and reproduction, confined by the iron law of rewriting to a mathematical analysis of similarity and difference whose boundary function will ultimately fail.

Voces clamantes

The comparison of these three texts raises the issue of whether one can define a relationship between the gender of the authors and their choice of desert landscape for translation. Similarities in theme and style between these works suggest that desert landscape may "speak" somewhat more to women than to men, and that women feel themselves called upon to be its translators for phallogocentric culture, in fact, to counter the very name of "desert" with notions of its peopling and aliveness. Annette Kolodny gives Appleton's refuge/prospect theory a feminist dimension by positing that the American landscape has functioned as a maternal symbol for males. The desert would seem to lack many of those features—sustenance, water, life, for example—normally associated with the Body of the Mother. Impossible to conceive of the desert "as woman, the total feminine principle of gratification--enclosing the individual in an environment of receptivity, repose, and painless and integral satisfaction" (*Lay of the Land* 4). In Kolodny's reading, the mental maps of generations of male American colonizers form into the single contour of a female breast: "the move to America was experienced as the daily reality of what has become its single dominating metaphor: regression from the cares of adult life and a return to the primal warmth of womb or breast in a feminine landscape" (6). The heart of darkness could be made palatable only when transformed through language and other signifying systems into a "harmony between man and nature based on an experience of the land as essentially feminine—that is, not simply the land as mother, but the land as woman, the total feminine principle of gratification—enclosing the individual in an environment of receptivity, repose, and painless and integral satisfaction" (6). Kolodny infers that the erotic possession of nature by man inevitably turns it into an urban landscape: "Colonization brought with it an inevitable paradox: the success of settlement depended on the ability to master the land, transforming the virgin territories into something else—a farm, a village, a road, a canal, a railway, a mine, a factory, a city, and finally, an urban nation. The instinctual drive embedded in the fantasy, which had first impelled men to emigrate, now impelled them both to continue pursuing the fantasy in daily life, and, when that failed, to codify it as part of the culture's shared dream life, through art" (7), or, alternatively, to relocate paradise farther West. Women's viewpoints, Kolodny argues, while not diametrically opposed to those of males, were not as easily summarized into a single, recuperative contour. The very word "paradise," for example, had different associations for women than for men: "The point here is not only that paradise implied different spheres of activity for men and for women, but it imaged qualitatively different spaces for those activities, also. For men it was a potential linguistic tool, with all its concomitant associations, for appropriating the vast extent of the new and unknown continent to human use (and, eventually, abuse). For women, it was from the first a term that denoted domesticity and the limited habitation of home. If the men altered the landscape to make it comply with their dreams of the pleasures, the bounty, and the receptivity of paradisal realms, the women patched quilts and embroidered bed hangings with brightly colored flowers and cultivated a nearby garden plot in order 'to render Home a Paradise'" ("Honing" 201). Women's quilts and hangings, then, are more than just decorative activities or accumulated labor. They constitute a complex, higher-level language describing the habitable space of their world. Many women's mental maps of America consist of language literally embroidered into a material cultivated in their own backyard. The extremely familial, heteroglossic, personal statement that emerges cannot be mapped onto the universalized male discourse of the virgin frontier. Prairie landscapes are dominated, according to Heather Avery, by the striking opposition between vertical and horizontal that imply man as conqueror or man as insignificant, either of which positions implies a thematic of culture versus nature (271). This literature has been dominated by male authors and male voices, but with some room for women authors, as represented for example by Martha Ostenso's *Wild Geese* (1925) and Gabrielle Roy's *La Route de Altamont* (1966).

Like the prairie, the desert provides an image of vastness and openness, of an immense and unlimited horizon dwarfing human verticality, but lacking the prairie's fertility and economic usefulness. Stephen Strom tries to capture this horizon in his photographs, on which are drawn the water trails and cloud patterns of Austin's descriptions; these photographs seem the proper canvas for tracing Mélanie's constant and precipitous wanderings in Brossard's fiction. The desert is denuded of the Baudelairean "forêt de symboles" that turns absolute velocity and becoming into eternal recurrence of the same. But just as gender relations are not symmetrical, neither can desert and forest oppose each other. The desert is not a father figure for these authors, for example; their relationship to it is not Oedipal. They develop not a counter-image, but a new way of seeing, mapping, and understanding, a fusion of horizons that is always an act of translation. Landscape in Austin, Brossard, and Harjo possesses the quality of *Übersetzbarkeit* ("translatability"), which Walter Benjamin defines as "a specific significance inherent in the original. . . . By virtue of its translatability the original is closely connected with the translation" (76). Translatability erases boundaries, of the kind erected by some other mental maps in this study. These are all texts of "lost borders."

Conclusion

By examining a series of fictions through a chronological sequence, this book has shown how canonical literary texts from a number of different languages and cultures of Europe and the Americas contributed, through their drawing of mental maps, to both the construction and contestation of national identity. As verbal maps of territorial, societal, and cultural boundaries, these texts stand in complex relationships to the problems of identity and nationalism prevalent at their time of writing. The balance of myth (the imaginary) and reality in these texts can be arranged along a continuum, from the Dikanka stories of Gogol that invoke an animistic world-view, to the deterministic environmentalism of Euclides da Cunha. Conversely, the role played by (social) scientific discourses and *mathesis* in the general mix of public perceptions of landscape and territory confronted and reworked by these authors vary from their direct usage by Goethe, Da Cunha, Benet, through their indirect influence on Arguedas, Faulkner, and Turgenev, to their suppression in favor of storytelling in Gogol, Harjo, and Brossard.

The relation of literary heterotopia to nation changes as we move chronologically from Goethe to Brossard. It is useful to think of four main groups: the eighteenth-century prenational or cultural-national stance of Goethe; the nineteenth-century focus on frontiers of the Other to construct nation in Gogol, Turgenev, and Da Cunha; the post-WWI regionalist response to nation in Faulkner, Lins do Rego, and Benet; and the reterritorialized landscapes of Arguedas, Brinkmann, Brossard, and Harjo, which find their precursors in the landscape lines of Mary Austin. From the fateful parallels of Goethe's texts and quasi-colonialist expansionism of Turgenev we moved to the tortured contradictions of Euclides da Cunha and the psychologized social landscapes of Faulkner, Lins do Rego, and Benet, to the self-reflexive identification of diataxis with translation in Austin, Harjo, and Brossard.

The differences between my use of heterotopia in this book and McHale's, as noted in the introduction, are instructive. McHale finds the term useful in characterizing postmodernist fiction. The Comala in Juan Rulfo's *Pedro Páramo*, the Impossible Mountains in Salman Rushdie's *Shame*, the Town in Haruki Murakami's *Hard-boiled Wonderland and the End of the World*, all represent heterotopias in McHale's sense of the term. The works I have treated, on the other hand, generally belong to traditions other than the postmodern, and the practice of literary realism influences many of them. The question here is whether postmodernism and whatever came after it represented a radical break in the uses of fictional mapping, particularly in its relationship to ideas of nation. I mention two developments as crucial in altering the role of fiction in the portrayal of national spaces, in fulfilling Ottmar Ette's prophecy that our study of contemporary literature must see it under the lens of "literature in motion." The first is the political fact of the writer-in-exile, a phenomenon that itself reflects the even larger political reality of displaced persons in the world today. The total number of refugees in the year 2006 was estimated at 20.8 million. (United Nations n.p.). Afghanis, Iraquis, Palestinians, Somalians, Sudanese, Palestinians — the list of groups that have been forced to relocate, to spend their lives in the twilit heterotopia of uncertain national status, seems to grow daily (in 1990 it was "only" fifteen million). If Rowan Oak provided us with the quintessential image of literary hermitage that allowed William Faulkner to draw a detailed geography of his region, the London and New York of Salman Rushdie represent the unhomely haunts of the quintessential, transnational modern writer-in-exile, who then writes about other, less cosmopolitan spaces. Swati Chanda, Marjorie Salvodon, Wendy Walters, and others have studied these maps of exile. In the introduction to her dissertation, Walters, like Ette, explicitly contrasts a literature of dwelling with a literature of movement: "This study [of African diasporic writers who do not live in their native countries] is part of a project of comparative cultural studies which seeks to intervene in this geographic ordering and instead privileges a nexus of movement over one of stasis, migration over dwelling" (1). Dwelling has indeed been both the product and the definition of the frontiers of nation. Goethe's state-sponsored vacation in Italy in no way resembles the displacement that has become a condition of existence for millions of people in many parts of the world.

Layered on top of these political realities is the theme of technological advances that allow instantaneous communication between various parts of the globe. It is possible that these advances, particularly space flight that allows a view of the whole earth, have worked what Carl Schmitt termed a "Raumrevolution," (spatial revolution) where "not only the measures and tools for measuring, not only the furthest horizon of mankind, but also the concept of space itself changes" (57). Taking up Schmitt's theme a generation later, and noting the technological changes Schmitt did not live to witness, Paul Virilio sees the "world-wide phenomenon of terrestrial and technological contraction that today makes us penetrate into an artificial topological universe: *the direct encounter of every surface on the globe*," adding further that beginning with the Cold War, "*geographic localization* seems to have definitively lost its strategic value and . . . this same value is attributed *to the delocalization of the vector*, of a vector in permanent movement" (47). Just how do we draw a map, mental or chorographic, where every surface encounters every other surface? And can we create maps that convey the motion of vectors? Mélanie's Meteor may be one such attempt. Joshua Meyrowitz argues that the simultaneity of events created by media such as television increasingly weakens the boundaries of class and nation, creating a neo-nomadic situation. Instantaneous communication produces "space-time compression" (Kellner). The different chronotope, as represented in the Internet, in turn evokes McHale's heterotopia, a simultaneous collocation of incommensurate "sites." Clearly, in a world of refugees national space becomes discontinuous, and the tools of the writer become deconstructive ones in line with McHale's definition of heterotopia. It is perhaps useful to compare these discontinuous sites of refugees to the prechorographic forms of mapping, which devalued spatial relations in favor of those of identification, usefulness, familiarity, sacrality, and so on. Globalization begets "glocalization." We return, as Doreen Massey points out, to a sense of place, but "what gives a place its specificity is not some long internalized history but the fact that it is constructed out of a particular constellation of social relations, meeting and weaving together at a particular locus" ("Global" 32). Landscape as a totalizing vision, the gaze of the landlord, has dissolved into a number of nonperspectival "-scapes," to use Arjun Appadurai's terminology.

All these factors contribute to a process that Gilles Deleuze and Félix Guattari have called "deterritorialization," a concept that Roger Rouse and Nikos Papastergiardis have applied to actual situations of migration and diaspora in different parts of the world. Ironically, the origins of deterritorialization as a "normal" way of life affecting large numbers of people coincide with the process of industrialization and urbanization that were occurring in England even as Goethe was visiting Italy. The urban landscape is constructed as an interplay of various social and symbolic forces analogous to what Gilles Deleuze call "smooth spaces," a concept I have applied to the sea in the work of Arguedas, and to the desert in the work of Austin, Harjo, and Brossard. A smooth space, associated with nomadic thought, is an open space that never closes upon itself. A striated space is defined by a central perspective, as in a landscape painting, by lines of latitude such as Goethe was preoccupied with, and by well-defined linguistic boundaries. A smooth space, on the other hand, is characterized by continual variation in orientations, landmarks, and linkages. Emily Apter, for example, has predicted the future as a "Deleuzian worldscape."

Finally, to displacement, simultaneity, and deterritorialization we must add the effect of virtuality. Some with training in biology argue seriously for the eventual possibility of a so-called "up-loading of consciousness," that is, the transfer of an individual's brain circuitry into a computer network, where it may "live" forever in a virtual digital environment (see the Transhumanism website at <http://www.aleph.se/Trans/>). This situation would be a fulfillment of Baudrillard's precession of simulacra, where the map literally pre-exists physical space, which in a cyberspace environment is zero-dimensional. From another direction, miniaturization may lead to what Virilio calls "the town in the body": "In the future, we will have the possibility of a technological colonization of the human body just as the geographical world was colonized by transports, communication, equipment. . . . What enabled the development of territories, towns and also the urban development will be applied to the human body just as if we had the town in the body and not only around the body. The town at home: *in vitro*, *in vivo*, the town in oneself" (Der Derian 68). Murakami's *Hard-Boiled Wonderland* represents this situation. Preceding the book is a map of The Town, which it turns out is essentially a computer program implanted in the protagonist's brain. Everything that happens in the town, and all its spaces, is virtual. One could speculate that the construction of the "town in the body," which is essentially a more intensive version of what is already happening with GPS systems that cyborgize their users, may affect our notion of mapping, turning us away from the subject-object dichotomy and geometric reduction that characterize chorographic maps, in favor of the map as an ergodic instrument — that is, we no longer see the map, but it makes our bodies move and guides our actions. In any case, there are clues that we may need to look to the body for some of the future true imaginary places of literature. An example that combines many of the observations made above into a single image is the face of Saleem, the protagonist of Salman Rushdie's *Midnight's Children*. Saleem's face is said to resemble the map of India. It is "read" by his teacher, Zagallo, and by his classmates: "'Thees is human geography! In the face of thees ugly ape don't you see the whole map of India?' 'Yes sir no sir you show us sir!' 'See here [Saleem's large nose]—the Deccan peninsula hanging down! . . . These stains . . . are Pakistan! Thees birthmark on the right ear is the East Wing, and thees horrible stained left cheek, the West!' . . . 'Lookit that, sir! The drip from his nose, sir! Is that supposed to be *Ceylon*?'" (277). Sure enough, to follow Deleuze's and Guattari's conceptual framework, the deterritorializing process of colonialism has made its subjects into a Body without Organs, for which Saleem's face stands. This passage might be compared with Gogol's anagogic description of the Dnieper River, in which natural features become part of a living, organic body. Here, in a reversal of the anagogic process, the living body of Saleem is reduced to chorographic political delineation. Nation and region, the text seems to say, inscribe themselves on the bodies of their citizens and inhabitants—in a process that results in deformity. Ominously, more than one nation can be discerned in Saleem's face, which we may thus consider to be at war with itself. Whether in the technological mode, as in Murakami, or the allegorical mode, as in Rushdie, literature predicts new senses of place. In addition to mental maps, there will be mind-body maps.

None of these developments has anything to do with the predicted withering of the nation-state. Along with Apter, Michael Hardt and Antonio Negri have produced typical overstatements of the case such as the following: "It is certainly true that, in step with the processes of globalization, the sovereignty of nation-states, while still effective, has progressively declined. The primary factors of production and exchange—money, technology, people, and goods—move with increasing ease across national boundaries; hence the nation-state has less and less power to regulate these flows and impose its authority over the economy. Even the most dominant nation-states should no longer be thought of as supreme and sovereign authorities, either outside or even within their own borders" (xi). Such predictions reflect the European and North American blinders that have been worn for centuries, although now such blinders eliminate rather than fortify concepts of nationalism and the nation-state. For example, the idea of simultaneity is far more apparent to those who possess the technology to make it happen than to those who still must travel by boat or on foot. The idea of transnational corporate data havens—the ultimate example of non-state economies, since any physical product will be subject to state regulation and surveillance—will not raise much of a response in the billions of people worldwide without data worth harvesting or stealing. Sovereign nation-states remain the chief drivers behind processes of globalization. Hardt and Negri's predictions seem based largely on the experiment of the European Union, and they ignore the fact that the majority of nations in existence today were created after 1945, and will probably continue their process of formation before they wither away. Rather than "*even* the most dominant nation-states," the last sentence of the above quote should perhaps read "*only* the most dominant nation-states." For example, the El Salvadoran scholar Silvia L. López can write in detail of the important role literature plays in "the process of construction of national culture" in her country after the end of its civil war (84). The end of the European colonial period created more new nation-states than any comparable period in history. Each of these has the potential to contribute its own set of mental maps, literary and otherwise, to the set considered here.

My purpose in concluding, then, is not to predict the demise either of mental maps or of nations, but rather to join Ottmar Ette in comparative speculation upon the interdisciplinary paradigms that will be needed to study whatever true imaginary places the future may create for us: "The dream of modernity, the creation of an increasingly homogenized space before the background of world traffic, world trade, world economy, has, to a large extent, become true. . . . But simultaneously with this development that has been quickening since the end of the 18th century, new cultural spaces have been developed, which do not stand for disdifferentiation but for new phenomena of differentiation. One might therefore assume an ongoing simultaneousness of both developments. That is why the problem of processes configuring space seems to me of such decisive meaning, not least for the literatures that are unfolding right now and that will be written in future, and the studies that are focused on them and try to learn from them" (12). Goethe with a GPS receiver?

Works Cited

Abrams, Robert E. *Landscape and Ideology in American Renaissance Literature.* Cambridge: Cambridge UP, 2004.

Adam, Wolfgang. "Arkadien als Vorhölle: Die Destruktion des Traditionellen Italien-Bildes in Rolf Dieter Brinkmanns *Rom, Blicke.*" *Euphorion* 83.2 (1989): 226-45.

Adams, Richard. *Faulkner: Myth and Motion.* Princeton: Princeton UP, 1968.

Adler, Hans. "Mediating the Mediation: Rolf Dieter Brinkmann's Concept of the Aesthetic Subject." *High and Low Cultures.* Ed. Reinhold Grimm and Jost Hermand. Madison: U of Wisconsin P, 1994. 95-105.

Agnew, John A., and James S. Duncan, eds. *The Power of Place.* Boston: Unwin Hyman, 1989.

Aiken, Charles S. "Faulkner's Yoknapatawpha County: Geographical Fact into Fiction." *The Geographical Review* 67.1 (1977): 1-21.

Alegría, Fernando. *Historia de la novela hispanoamericana.* Mexico: Siglo Veintuno, 1974.

Almeida Prado, Antônio Lázaro de. "Universo em recesso: no fogo morto . . . da fala." *Revista de letras* 16 (1974): 47-72.

Althaus, Horst. *Ästhetik, Ökonomie und Gesellschaft.* Bern: Francke, 1971.

Amodea, Immacolata. "Rolf Deiter Brinkmanns Versuch, ohne Goethe über Italien zu Schreiben." *arcadia* 34.1 (1999): 2-19.

Anderson, Benedict. *Imagined Communities: Reflections on the Origin and Spread of Nationalism.* London: Verso, 1991.

Andrade Tosta, Antônio Luciano de. "O 'sublime' sertão em *O Sertanejo*, *Os sertões* e *Grande Sertão: Veredas.*" *Letras de hoje* 36.1 (2001): 7-26.

Andrade, Mário de. "'Fogo morto.'" *O Empalhador de pássarinho. Obras completas.* Vol. 20. São Paulo: Martins, 1972. 291-95.

Andrade, Mário de. "Lasar Segall." *Aspectos das artes plásticas no Brasil. Obras completas.* Vol. 12. São Paulo: Martins, 1965. 47-67.

Appadurai, Arjun. *Modernity at Large.* Minneapolis: U of Minnesota P, 1996.

Appleton, Jay. *The Experience of Landscape.* Rev. ed. Chichester: Wiley, 1996.

Apter, Emily. "Nomadologies of Tomorrow: The Deleuzian Worldscape." *Continental Drift: From National Characters to Virtual Subjects*. Chicago: U of Chicago P, 1999. 225-37.

Arguedas, José María. *Agua*. 1935. Lima: Dirección Universitaria de Biblioteca y Publicaciones, 1974.

Arguedas, José María. "Agua." *Agua*. Lima: Dirección Universitaria de Biblioteca y Publicaciones, 1974. 1-46.

Arguedas, José María. *Deep Rivers*. Trans. Frances H. Barraclough. Austin: U of Texas P, 1978.

Arguedas, José María. *El zorro de arriba y el zorro de abajo*. 1971. *Obras completas*. Ed. Sybilia Arrendondo de Arguedas. Vol. 5. Lima: Horizonte, 1983. 9-219.

Arguedas, José María. "Los escoleros." *Agua* (1935): 47-98.

Arguedas, José María. *Los rios profundos*. Buenos Aires: Losada, 1958.

Arguedas, José María. "Orovilca." *La agonía de Rasu-Ñiti*. Ed. Abelardo Oquendo. Lima: Populibros Peruanos, n.d. 74-97.

Arguedas, José María. "Sabogal y las artes populares en el Perú." *Folklore Americano* 4.4 (1956): 241-45.

Arguedas, José María. *Todas las sangres*. *Yawar fiesta*. *Obras completas*. Ed. Sybilia Arrendondo de Arguedas. Vol. 2. Lima: Horizonte, 1983.

Arguedas, José María. *Yawar Fiesta*. Trans. Frances H. Barraclough. Austin: U of Texas P, 1985.

Arguedas, José María. "Yo soy hechura de mi madrastra." *Páginas Escogidas*. Lima: Universo, 1972. 247-53.

Atwood, Margaret. *Strange Things: The Malevolent North in Canadian Literature*. Oxford: Clarendon, 1995.

Austin, Mary. *The American Rhythm*. New York: Harcourt-Brace, 1923.

Austin, Mary. *The Land of Little Rain*. 1902. Boston: Houghton Mifflin, 1950.

Austin, Mary. "The Lost Mine of Fisherman's Peak." *Western Trails*. Ed. Melody Graulich. Reno: U of Nevada P, 1987. 209-23.

Austin, Mary. *Lost Borders*. 1909. *Stories from the Country of Lost Borders*. Ed. Marjorie Pryse. New Brunswick: Rutgers, 1987.

Austin, Mary. "Non-English Writing II." *Cambridge History of American Literature*. Ed. William P. Trent, John Erskine, Stuart P. Sherman, and Carl Van Doren. Vol. 3. New York: MacMillan, 1961. 610-34.

Bahloul, Joëlle. *The Architecture of Memory*. Cambridge: Cambridge UP, 1996.

Bakhtin, Mikhail. "The *Bildungsroman* and Its Significance in the History of Realism (Toward a Historical Typology of the Novel)." Trans. Vern W. McGee. *Speech Genres and other Late Essays*. Ed. Carly Emerson and Michael Holquist. Austin: U of Texas P, 1986. 10-59.

Bamyeh, Mohammed A. *The Ends of Globalization*. Minneapolis: U of Minnesota P, 2000.

Bárbero, Alessandro. *Bella vita e guerre altrui de Mr. Pyle, gentiluomo*. Milano: Mondadori, 1995.

Barnes, Trevor J., and James S. Duncan. *Writing Worlds: Discourse, Text and Metaphor in the Representation of Landscape*. New York: Routledge, 1992.

Bassin, Mark. "Russia between Europe and Asia: The Ideological Construction of Geographical Space." *Slavic Review* 50 (1991): 1-17.

Bassin, Mark. "Russian Geographers in the Far East." *Geography and National Identity*. Ed. David Hooson. Oxford: Blackwell, 1994. 112-33.

Basso, Keith. "Wisdom Sits in Places: Notes on a Western Apache Landscape." *Senses of Place*. Ed. Steven Feld and Keith H. Basso. Santa Fe: School of American Research P, 1996. 53-90.

Battafarano, Italo Michele. *Die im Chaos blühenden Zitronen*. Bern: Peter Lang, 1999.

Baudrillard, Jean. *America*. Trans. Chris Turner. London: Verso, 1988.

Bauer, Ralph. *The Cultural Geography of Colonial America*. New York: Cambridge UP, 2003.

Beller, Manfred. "Geschichtserfahrung und Selbstbespiegelung im Deutschland-Bild der Italienischen und im Italien-Bild der deutschen Gegenwartsliteratur." *arcadia* 17 (1982): 154-70.

Benet, Juan. "Breve historia de *Volverás a Región*." *Revista de Occidente* 45.134 (1974): 160-65.

Benet, Juan. "De Canudos a Macondo." *Revista de Occidente* 24.70 (1969): 49-57.

Benet, Juan. "El Agua en Región." *La Moviola de Euripides*. Madrid: Taurus, 1981. 67-75.

Benet, Juan. *El aire de un crimen*. Barcelona: Planeta, 1980.

Benet, Juan. *Herrumbrosas lanzas*. Madrid: Alfaguara, 1983-86.

Benet, Juan. "Numa, una leyenda." *Cuentos completos*. 2nd ed. Ed. Juan Benet. Vol. 1. Madrid: Alianza, 1981. 237-82.

Benet, Juan. *La otra casa de Mazón*. Barcelona: Seix Barral, 1973.

Benet, Juan. "Región." *Páginas impares*. Buenos Aires: Alfaguara, 1996. 93-98.

Benet, Juan. *Return to Region*. Trans. Gregory Rabassa. New York: Columbia UP, 1985.

Benet, Juan. *Una meditación*. Madrid: Alfaguara, 1985.

Benet, Juan. *Volverás a Región*. Madrid: Destino, 1967.

Benítez-Rojo, Antonio. "From the Plantation to the Plantation." *The Repeating Island*. Trans. James E. Maraniss. Durham: Duke UP, 1992. 33-81.

Benjamin, Walter. "The Storyteller." *Illuminations: Essays and Reflections*. Ed. Hannah Arendt. New York: Schocken, 1978. 83-109.

Benjamin, Walter. "The Task of the Translator." Trans. Harry Zohn. *The Translation Studies Reader*. Ed. Lawrence Venuti. New York: Routledge, 2000. 75-85.

Berger, John. *Ways of Seeing*. London: Penguin, 1972.

Berman, Marshall. *All that is Solid Melts into Air*. New York: Simon & Schuster, 1982.

Bernstein, Susan. "Goethe's Architectonic Bildung and Buildings in Classical Weimar." *Modern Language Notes* 114.5 (1999): 1014-36.

Bhabha, Homi. "DissemiNation." *Nation and Narration* 291-322.

Bhabha, Homi. "Introduction: Narrating the Nation." *Nation and Narration*.1-7.

Bhabha, Homi, ed. *Nation and Narration*. New York: Routledge, 1990.

Black, David, Donald Kunze, and John Pickles. *Commonplaces: Essays on the Nature of Place*. New York: UP of America, 1989.

Bode, Wilhelm, ed. *Goethe in vertraulichen Briefen seiner Zeitgenossen*. München: Beck, 1982.

Boelhower, William. *Through a Glass Darkly: Ethnic Semiosis in American Literature*. New York: Oxford UP, 1987.

Boerner, Peter. "Italienische Reise (1816-29)." *Interpretationen. Goethes Erzählwerk*. Stuttgart: Reclam, 1985. 344-62.

Bohannon, Paul. "Space and Territoriality." *Africa and Africans*. Garden City: Natural History Press, 1964.

Bordessa, Ronald. "The City in Canadian Literature: Realist and Symbolic Interpretations." *The Canadian Geographer / Le Géographe canadien* 32 (1988): 272-76.

Borges, Jorge Luis. "The Garden of Forking Paths." *Labyrinths*. Trans. Donald A. Yates. New York: New Directions, 1964. 19-29.

Borges, Jorge Luis. "El jardín de senderos que se bifurcan." *Ficciones*. Buenos Aires: Emece, 1956. 11-59.

Brinkmann, Rolf Dieter. *Rom, Blicke*. Reinbek bei Hamburg: Rowohlt, 1979.

Brooks, Peter. *Reading for the Plot*. Cambridge: Harvard UP, 1984.

Brossard, Nicole. *Le Désert mauve*. Québec: Hexagone, 1987.

Brossard, Nicole. "Nicole Brossard." Trans. Susanne de Lotbinière-Harwood. *Contemporary Authors Autobiography Series*. Ed. Hal May and Susan M. Trotsky. Detroit: Gail Research Company, 1993. 39-57.

Brossard, Nicole. *Mauve Desert*. Trans. Susanne de Lotbinière-Harwood. Toronto: Coach House P, 1990.

Brosseau, Marc. *Des Romans-géographes. Essai*. Paris: L'Harmattan, 1996.

Brower, Daniel, and Edward Lazzerini, eds. *Russia's Orient: Imperial Borderlands and Peoples, 1700-1917*. Bloomington: Indiana UP, 1997.

Brown, Calvin S. "Faulkner's Geography and Topography." *Publications of the Modern Language Association of America* 77 (1962): 652-59.

Burán, M. "Juan Benet y la nueva novela española." *Cuadernos Americanos* 183 (1974): 193-205.

Burns, Margie. "A Good Rose is Hard to Find." *Image and Ideology in Modern/Postmodern Discourse*. Ed. David B. Downing and Susan Bazargan. Albany: State U of New York P, 1991. 105-23.

Buttimer, Anne, ed. *The Practice of Geography*. New York: Longman, 1983.

"California: Adopt-A-Desert." *Sierra* (July-August 2001): 67-68.

Casey, Edward. *The Fate of Place: A Philosophical History*. Berkeley: U of California P, 1997.

Casey, Edward. *Getting Back Into Place*. Bloomington: Indiana UP, 1993.

Castello, José Aderaldo. "Memória, primitivismo e regionalismo." *José Lins do Rego*. Ed. Eduardo F. Coutinho and Ângela Bezerra de Castro. Rio de Janeiro: FUNESC, 1990. 183-89.

Castro-Klarén, Sara. *El mundo mágico de José María Arguedas*. Lima: Instituto de Estudios Hispanoamericanos, 1973.

Certeau, Michel de. *The Practice of Everyday Life*. Berkeley: U of California P, 1984.

Certeau, Michel de. "Practices of Space." *On Signs*. Ed. Marshall Blonsky. Baltimore: Johns Hopkins UP, 1985. 122-45.

Chanady, Amaryll. "Introduction: Latin American Imagined Communities and the Postmodern Challenge." *Latin American Identity and Constructions of Difference*. Ed. Amaryll Chanady. Minneapolis: U of Minnesota P, 1994. ix-xlvi.

Chanda, Swati. *Narratives of Nation in the Age of Diaspora*. Ph.D. diss. West Lafayette: Purdue U, 1996.

Chard, Chloe. *Pleasure and Guilt on the Grand Tour*. Manchester: Manchester UP, 1999.

Chatwin, Bruce. *The Songlines*. London: Jonathan Cape, 1987.

Coelho Fontes, Oleone. *O Treme-Terra: Moreira César, a República, e Canudos*. Salvador: Vozes, 1995.

Cohn, Deborah. "Faulkner and Spanish America: Then and Now." *Faulkner in the Twenty-First Century*. Ed. Robert W. Hamblin and Ann J. Abadie. Jackson: UP of Mississippi, 2003.

Cohn, Deborah. *History and Memory in the Two Souths*. Nashville: Vanderbilt UP, 1999.

Collins, Carvel. "Christian and Freudian Structures." *Twentieth-Century Interpretations of* The Sound and the Fury. Ed. Michael H. Cowan. Englewood Cliffs: Prentice-Hall, 1968. 73-74.

Compitello, Malcolm A. *Ordering the Evidence:* Volverás a Región *and Civil War Fiction*. Barcelona: Puvill, 1983.

Compitello, Malcolm A. "Region's Brazilian Backlands." *Hispanic Journal* 1.2 (1980): 25-45.

Conley, Tom. *The Self-Made Map: Cartographic Writing in Early Modern France*. Minneapolis: U of Minnesota P, 1996.

Cornejo Polar, Antonio. *Los universos narrativos de José Maria Arguedas*. Buenos Aires: Losada, 1973.

Corrêa, Nereu. "A tapeçaria lingüística d'*Os Sertões*." *A tapeçaria lingüística d'*Os Sertões *e outros estudos*. By Nereu Corrêa. São Paulo: Edições Quirón, 1978. 1-28.

Cosgrove, Denis. *Social Formation and Symbolic Landscape*. London: Croom Helm, 1984.

Costa Lima, Luiz. *Control of the Imaginary*. Trans. Ronald W. Sousa. Minneapolis: U of Minnesota P, 1988.

Coutinho, Eduardo F. "A relação arte/realidade em *Fogo morto*." *José Lins do Rego*. Ed. Eduardo F. Coutinho and Ângela Bezerra de Castro. João Pessoa: Funesc, 1990. 430-40.

Cruz Leal, Petra Iraides. "*Los rios profundos*: Realidad científica y proyección estética." *Alba de América* 8.14-15 (1990): 253-62.

Da Cunha, Euclides. *Canudos: Diaria de uma expedicão*. Rio de Janeiro: José Olympio, 1939.

Da Cunha, Euclides. *Os Sertões* 1902. Rio de Janeiro: Francisco Alves, 1982.

Da Cunha, Euclides. *Rebellion in the Backlands*. Trans. Samuel Putnam. Chicago: U of Chicago P, 1944.

Dantas, Paulo. *Os Sertões de Euclides e outros Sertões*. São Paulo: Conselho Estadual de Cultura, 1969.

Davis, Thadious M. *Faulkner's "Negro."* Baton Rouge: Louisiana State UP, 1983.

De Man, Paul. *Allegories of Reading*. New Haven: Yale UP, 1979.

Deleuze, Gilles, and Félix Guattari. *A Thousand Plateaus*. Trans. Brian Massumi. Minneapolis: U of Minnesota P, 1987.

Der Derian, James. *Virtuous War*. Boulder: Westview P, 2001.

Desvergnes, Alain. *Yoknapatawpha: The Land of William Faulkner*. Paris: Marval, 1990.

Diprose, Rosalyn, and Robyn Ferrell, eds. *Cartographies: Poststructuralism and the Mapping of Bodies and Spaces*. Sydney: Allen and Unwin, 1991.

Dixon, Melvin. *Ride Out the Wilderness: Geography and Identity in Afro-American Literature*. Urbana: U of Illinois P, 1987.

Draper, R. P., ed. *The Literature of Region and Nation*. London: MacMillan, 1989.

Durand, Regis. "'This Shadowy Attenuation of Time....'" *Yoknapatawpha: The Land of William Faulkner*. By Alain Desvergnes. Paris: Marval, 1990. n.p.

During, Simon, ed. *Cultural Studies: A Critical Introduction*. New York: Routledge, 2005.

Durkin, Andrew R. "Turgenev and the Pastoral Tradition." *American Contributions to the Eleventh International Congress of Slavists, Bratislava, August-September 1993: Literature, Linguistics, Poetics*. Ed. Robert Maguire and Alan Timberlake. Columbus: Slavica, 1993. 43-50.

Edney, Matthew H. *Mapping an Empire: The Geographical Construction of British India, 1765-1843*. Chicago: U of Chicago P, 1997.

Ellison, Fred. *Brazil's New Novel*. Berkeley: U of California P, 1954.

Entrikin, Nicholas. "Geography's Spatial Perspective and the Philosophy of Ernst Cassirer." *Canadian Geographer / Le Géographe canadien* 21 (1977): 209-22.

Entrikin, Nicholas. *The Betweenness of Place.* Baltimore: Johns Hopkins UP, 1991.

Erlich, Victor. *Gogol.* New Haven: Yale UP, 1969.

Ette, Ottmar. *Literature on the Move.* Trans. Katharina Vester. Amsterdam: Rodopi, 2003.

Eyles, John. *Senses of Place.* Warrington: Silverbrook P, 1985.

Fanger, Donald. *The Creation of Nikolai Gogol.* Cambridge: Harvard UP, 1979.

Fanger, Donald. "Dead Souls: The Mirror and the Road." *Nineteenth-Century Fiction* 33 (1978): 24-47.

Faulkner, William. *Absalom, Absalom!* New York: Viking, 1936.

Faulkner, William. *Le Bruit et la fureur.* Trans. Maurice Coindreau. Paris: Gallimard, 1938.

Faulkner, William. *The Hamlet.* New York: Random House-Vintage, 1958.

Faulkner, William. "Mississippi." *Essays, Speeches and Public Letters.* Ed. James B. Meriwether. New York: Random House, 1966. 11-43.

Faulkner, William. *Sanctuary.* New York: Random House, 1931.

Faulkner, William. *Sartoris.* New York: Harcourt Brace, n.d.

Faulkner, William. *O som e a fúria.* Trans. Mario Henrique Leiria and H. Santos Carvalho. Ed. Luis de Sousa Rebelo. Lisbon: Portugalia, 1960.

Faulkner, William. *The Sound and the Fury.* 1929. New York: Vintage-Random House, 1987.

Feld, Steven, and Keith H. Basso, eds. *Senses of Place.* Santa Fe: School of American Research P, 1996.

Foote, Shelby. "Faulkner's Depiction of the Planter Aristocracy." *The South and Faulkner's Yoknapatawpha.* Ed. Evans Harrington and Ann J. Abadie. Jackson: U of Mississippi P, 1977. 40-61.

Foucault, Michel. "Of Other Spaces." Trans. Jay Miskowiec. *Diacritics* 16.1 (1986): 22-27.

Foucault, Michel. "Space, Power, Knowledge." Trans. Christian Hubert. *The Cultural Studies Reader.* Ed. Simon During. New York: Routledge, 1993. 151-61.

Fowler, Doreen. "'Little Sister Death': *The Sound and the Fury* and the Denied Unconscious." *Faulkner and Psychology.* Ed. Donald Kartiganer and Ann J. Abadie. Jackson: U of Mississippi P, 1994. 3-20.

Freyre, Gilberto."Introduction." *Canudos: Diario de uma expedição.* By Euclides da Cunha. Ed. Antônio Simões dos Reis. Rio de Janeiro: José Olympio, 1939. vii-xxv.

Freyre, Gilberto. "O Pai e o filho." *Sobrados e mucambos.* Vol. 1. Rio de Jañeiro: José Olympio, 1981. 67-92.

Freyre, Gilberto. Unpublished letter to José Lins do Rego, 29 November 1938. João Pessoa: Espaço Cultural José Lins do Rego.

Furst, Lilian "Goethe's *Italienische Reise* in its European Context." *Goethe in Italy, 1786-1986.* Ed. Gerhart Hoffmeister. Amsterdam: Rodopi, 1988. 115-32.

Fusso, Susanne. "The Landscape of Arabesques." *Essays on Gogol: Logos and the Russian Word.* Ed. Susanne Fusso and Priscilla Meyer. Evanston: Northwestern UP, 1992. 112-25.

Gadamer, Hans-Georg. *Wahrheit und Methode.* Tübingen: Mohr, 1972.

Gama e Melo, Virgínius da. *Estudos críticos.* João Pessoa: Editora Universitária, 1980.

García Márquez, Gabriel. "William Faulkner 1897/1997." *A Faulkner 100: The Centennial Exhibition. With a Contribution by Gabriel García Márquez.* Ed. Thomas Verich. Jackson: U of Mississippi Library Special Collections, 1997. n.p.

Gates, Henry Louis. *The Signifying Monkey.* New York: Oxford UP, 1988.

Gellner, Ernest. *Nations and Nationalism.* Ithaca: Cornell UP, 1983.

Genovese, Eugene. *The World the Slaveholders Made.* New York: Pantheon, 1969.

Gephart, Werner. "Goethe als 'Gesellschaftsforscher'? Eine soziologische Lektüre der *Italienischen Reise.*" *Goethe und Italien.* Ed. Willi Hirdt and Birgit Tappert. Bonn: Bouvier, 2001. 105-24.

Gerhard, Melitta. "Die Redaktion der *Italienischen Reise* im Lichte von Goethes autobiographischem Gesamtwerk." *Leben im Gesetz. Fünf Goethe-Aufsätze.* By Melitta Gerhard. Bern: Francke, 1966. 34-51.

Gillies, John. *Shakespeare and the Geography of Difference.* New York: Cambridge UP, 1994.

Glissant, Edouard. *Faulkner, Mississippi.* Trans. Barbara Lewis and Thomas C. Spear. New York: Farrar, Straus & Giroux, 1999.

Godden, Richard. "Quentin Compson: Tyrrhenian Vase of Crucible of Race?" *New Essays on* The Sound and the Fury. Ed. Noel Polk. Cambridge: Cambridge UP, 1993. 99-138.

Goethe, Johann Wolfgang von. "Einleitung zu Th. Carlyle. Leben Schillers." *Schriften zur Literatur.* Trans. Ellen von Nardroff and Ernest H. von Nardroff. Ed. Erich Trunz. München: Beck, 1981.

Goethe, Johann Wolfgang von. *The Flight to Italy: Diary and Selected Letters.* By Johann Wolfgang von Goethe. Trans. T.J. Reed. Oxford: Oxford UP, 1999.

Goethe, Johann Wofgang von. *Italian Journey.* Trans. Robert R. Heitner. Ed. Thomas P. Saine and Jeffrey L. Sammons. Frankfurt: Suhrkamp, 1989.

Goethe, Johann Wofgang von. *Italienische Reise.* 1816. By Johann Wolfgang von Goethe. Ed. Erich Trunz. München: Beck, 1981.

Goethe, Johann Wofgang von. "Literarischer Sanskulottismus." *Schriften zur Literatur.* Ed. Erich Trunz. München: Beck, 1981. 239-44.

Goethe, Johann Wofgang von. "Response to a Literary Rabble-Rouser." Trans. Ellen von Nardroff and Ernest H. von Nardroff. *Essays on Art and Literature.* Ed. John Gearey. Frankfurt: Suhrkamp, 1986. 189-92.

Goethe, Johann Wolfgang von. "Schicksal der Handschrift." *Zur Naturwissenschaft überhaupt, besonders zur Morphologie. Erfahrung, Betrachtung, Folgerung durch Lebensereignisse verbunden.* By Johann Wolfgang von Goethe. Ed. Hans J. Becker. München: Carl Hanser, 1989. 69-72.

Goethe, Johann Wolfgang von. *Tagebuch der italienischen Reise 1786.* Frankfurt: Insel, 1976.

Goethe, Johann Wolfgang von. "World Literature." *Essays on Art and Literature.* Ed. John Gearey. Frankfurt: Suhrkamp, 1986. 224-28.

Gogol, Nikolai. *Dead Souls.* Trans. Richard Pevear and Larissa Volokhonsky. New York: Pantheon, 1996.

Gogol, Nikolai. *Evenings on a Farm Near Dikanka.* 1831-32. *The Complete Tales of Nikolai Gogol.* Trans. Constance Garnett. Ed. Leonard J. Kent. Vol. 1. Chicago: U of Chicago P, 1985. 3-206.

Gogol, Nikolai. "Four Letters to Divers Persons." *Dead Souls.*Trans. and Ed. George Gibian. New York: Norton, 1985. 411-23.

Gogol, Nikolai. *Mertvye dushi.* 1842. *Sobranenie Socinenij.* By Nikolai Gogol. Moscow: Government Literary Publisher, 1953.

González Echevarria, Roberto. *Myth and Archive: A Theory of Latin American Narrative.* Cambridge: Cambridge UP,1990.

Goodman, Audrey. *Translating the American Southwest.* Tuscon: U of Arizona P, 2002.

Gould, Peter R., and Rodney White. *Mental Maps.* Boston: Allen & Unwin, 1986.

Grabowicz, George G. "Three Perspectives on the Cossack Past." *Harvard Ukrainian Studies* 5.2 (1981): 171-94.

Gran Atlas de España. Barcelona: Planeta, 1990.

Grandis, Rita de. "La mirada antropológica de Arguedas." *El Indio: nacimiento y evolución de una instancia discursiva.* Montpellier: Centre d'Études et de Recherches Sociocritiques, 1994. 109-24.

Greenfield, Gerald Michael. "*Sertão* and *Sertanejo*: An Interpretive Context for Canudos." *Luso-Brazilian Review* 30.2 (1993): 35-46.

Griffiths, Frederick T., and Stanley J. Rabinowitz. *Novel Epics: Gogol, Dostoevsky, and National Narrative.* Evanston: Northwestern UP, 1990.

Grimm, Gunter E., Ursula Breymayer, and Walter Erhart. "Entzauberung eines Mythos? Zum Italienbild deutscher Gegenwartsliteratur." *"Ein Gefühl von freierem Leben." Deutsche Dichter in Italien.* Ed. Gunter E. Grimm, Ursula Breymayer, and Walter Erhart. Stuttgart: Metzler, 1990. 283-301.

Grossman, Leonid. "Turgenev's Early Genre." *Critical Essays on Ivan Turgenev.* Ed. David A. Lowe. Boston: Hall, 1989. 63-73.

Guimarães Rosa, João. *Grande Sertão: Veredas.* Rio de Janeiro: Nova Fronteira, 1986.

Gullón, Ricardo. "Una región laberíntica que bien pudiera llamarse España." *Insula* 319 (1973): 3-10.

Hachmeister, Gretchen L. *Italy in the German Literary Imagination: Goethe's "Italian Journey" and its Reception by Eichendorff, Platen, and Heine.* Rochester: Camden House, 2002.

Häkli, Jouni. "Cultures of Demarcation: Territory and National Identity in Finland." *Nested Identities: Nationalism, Territory, and Scale*. Ed. Guntram H. Herb and David H. Kaplan. Lanham: Rowman & Littlefield, 1999. 123-49.

Hard, Gerhard. *Die "Landschaft" der Sprache und die "Landschaft" der Geographen*. Bonn: Dümmler, 1970.

Hardt, Michael, and Antonio Negri. *Empire*. Cambridge: Harvard UP, 2000.

Harjo, Joy, and Stephen Strom. *Secrets from the Center of the World*. Tucson: Sun Tracks & U of Arizona P, 1989.

Harley, J.B. "New England Cartography and the Native Americans." *American Beginnings*. Ed. Emerson W. Baker, Edwin A. Churchill, Richard S. D'Abate, Kristine L. Jones, Victor A. Konrad, and Harald E.L. Prins. Lincoln: U of Nebraska P, 1994. 287-313.

Harley, J.B. *Writing Worlds: Discourse, Text and Metaphor in the Representation of Landscape*. New York: Routledge, 1992.

Hassinger, Hugo. "Der Staat als Landschaftsgestalter." *Zeitschrift für Geopolitik* 9 (1932): 117-22, 182-87.

Heidegger, Martin. *Die Frage nach dem Ding*. Frankfurt: Vittorio Klostermann, 1984.

Herb, Guntram H., and David H. Kaplan, eds. *Nested Identites: Nationalism, Territory, and Scale*. Lanham: Rowman & Littlefield, 1999.

Hines, Thomas S. *William Faulkner and the Tangible Past: The Architecture of Yoknapatawpha*. Berkeley: U of California P, 1996.

Howell, Elmo. "William Faulkner and the Plain People of Yoknapatawpha County." *The Journal of Mississippi History* 24.2 (1962): 73-87.

Huggan, Graham. *Territorial Disputes: Maps and Mapping Strategies in Contemporary Canadian and Australian Fiction*. Toronto: U of Toronto P, 1994.

Hyart,Charles. "Le Dynamisme caché de la littérature russe." *Revue Générale* 4 (1981): 25-37.

Irwin, John. *The Mystery to a Solution*. Baltimore: Johns Hopkins UP, 1994.

Ivo, Lêdo. "Prefácio." *O vulcão e a fonte*. By José Lins do Rêgo. Rio de Janeiro: Cruzeiro, 1958.

Jacobson, Mark, and Scott Thode. "Down to the Crossroads." *Natural History* 105.9 (1996): 51-55.

Johnson, Richard. "What is Cultural Studies Anyway?" *Social Text* 16 (1986): 38-80.

Johnson, Robert. "Dust My Broom." *The Complete Recordings*. New York: Columbia Records 46222,

Jünger, Ernst and Carl Schmitt. *Briefe 1930-1983*. Ed. Helmuth Kiesel. Stuttgart: Klett-Cotta, 1999.

Justus, James H. "Faulkner's Fortunate Geography." *William Faulkner: The Yoknapatawpha Fiction*. Ed. A. Robert Lee. London: Vision P, 1990. 19-41.

Kandiyoti, D. "Identity and its Discontents: Women and the Nation." *Millenium: Journal of International Studies* 20.3 (1991): 429-43.

Kaplan, David H. "Territorial Identities and Geographic Scale." *Nested Identities*. Ed. Guntram H. Herb and David H. Kaplan. Lanham: Rowman & Littlefield, 1999.

Kartiganer, Donald, and Ann J. Abadie. *William Faulkner and the Natural World.* Jackson: U of Mississippi P, 1999.

Kellner, Douglas. "Globalization and the Postmodern Turn." (2001): <http://www.gseis.ucla.edu/courses/ed253a/dk/GLOBPM.htm>.

Kiefer, Klaus H. *Wiedergeburt und Neues Leben. Aspekte des Struckturwandels in Goethes Italienischer Reise*. Bonn: Herbert Grundmann, 1978.

Kirwan, Albert D. *Revolt of the Rednecks*. Lexington: U of Kentucky P, 1957.

Klenze, Camillo von. *The Interpretation of Italy During the Last Two Centuries*. Chicago: U of Chicago P, 1907.

Klimenko, Vladimir. "Zemlia Gogolia." *Literaturnoe obozrenie* 12 (1979): 86

Koepke, Wulf. "Goethe and the Aesthetic Education of the Germans." *Goethe as a Critic of Literature*. Ed. Karl J. Fink and Max L. Baeumer. New York: UP of America, 1984. 86-109.

Kolodny, Annette. "Honing a Habitable Languagescape: Women's Images for the New World Frontiers." *Women and Language in Literature and Society*. Ed. Sally McConnell-Ginet, Ruth Borker, and Nelly Furman. New York: Praeger, 1980. 188-204.

Kolodny, Annette. *The Land Before Her: Fantasy and Experience of the American Frontier, 1630-1860*. Chapel Hill: U of North Carolina P, 1984.

Kolodny, Annette. *The Lay of the Land: Metaphor as Experience and History in American Life*. Chapel Hill: U of North Carolina P, 1975.

Kopper, John. "The 'Thing-in-Itself' in Gogol's Aesthetics." *Essays on Gogol: Logos and the Russian Word.* Ed. Susanne Fusso and Priscilla Meyer. Evanston: Northwestern UP, 1992. 40-62.

Kornblatt, Judith Deutsch. *The Cossack Hero in Russian Literature*. Madison: U of Wisconsin P, 1992.

Koropeckyj, Roman, and Robert Romanchuk. "Ukraine in Blackface: Performance and Representation in Gogol's Dikan'ka Tales, Book One." *Slavic Review* 62.3 (2003): 525-47.

Kort, Wesley A. *Place and Space in Modern Fiction*. Gainesville: UP of Florida, 2004.

Kortenaar, Neil ten. "Fictive States and the State of Fiction in Africa." *Comparative Literature* 52.3 (2000): 228-44.

Krippendorf, Ekkehart. *Goethe. Politik gegen den Zeitgeist*. Frankfurt: Insel, 1999.

Kristeva, Julia. "The Father, Love, and Banishment." *Psychoanalysis and Literature*. Ed. Edith Kurzweil and William Phillips. New York: Columbia UP, 1983. 389-99.

Labanyi, Jo. "Fiction as Echo: *Volverás a Región.*" *Myth and History in the Contemporary Spanish Novel.* By Jo Labanyi. Cambridge: Cambridge UP, 1989. 95-134.

Lacoste, Jean. *"Le Voyage en Italie" de Goethe.* Paris: PU de France, 1999.

Ladd, Barbara. *Nationalism and the Color Line.* Baton Rouge: Louisiana State UP, 1996.

Landeira, António Luís. "O Nordeste e José Lins do Rego: Da desintegração de um sistema—ou ainda o Nordeste." *Vértice: Revista de Cultura e Arte* 30 (1970): 472-81.

Laplanche, Jean, and J.B. Pontalis. *The Language of Psycho-Analysis.* Trans. Donald Nicholson-Smith. New York: Norton, 1973.

Lawrence, John, and Dan Hise. *Faulkner's Rowan Oak.* Jackson: UP of Mississippi, 1993.

Layoun, Mary. *Wedded to the Land? Gender, Boundaries, and Nationalism in Crisis.* Durham: Duke UP, 2001.

Layton, Susan. *Russian Literature and Empire: Conquest of the Caucasus from Pushkin to Tolstoy.* Cambridge: Cambridge UP, 1994.

Lefebvre, Henri. *The Production of Space.* Trans. Donald Nicholson-Smith. Oxford: Blackwell, 1991.

Lehmann, Herbert. *Goethe und Gregorovius vor der italienischen Landschaft.* Wiesbaden: Steiner, 1967.

Lewin, Linda. *Politics and Parentela in Paraíba: A Case Study of Family-Based Oligarchy in Brazil.* Princeton: Princeton UP, 1987.

Ley, David, and Marwyn S. Samuels, eds. *Humanistic Geography: Prospects and Problems.* Chicago: Maaroufa, 1978.

Lins do Rego, José. "Augusto dos Anjos e o Engenho Pau d'Arco." *Homens, seres, e coisas.* By José Lins do Rego. Rio de Janeiro: Ministério da Educação e Saúde, 1957. 1-20.

Lins do Rego, José. *Bangüê.* 1934. Rio de Janeiro: José Olympio, 1966.

Lins do Rego, José. *Fogo morto.* 1942. Rio de Janeiro: José Olympio, 1983.

Lins do Rego, José. *Menino de Engenho.* 1932. Rio de Janeiro: José Olympio, 1980.

Lins do Rego, José. *Meus verdes anos.* 1956. Rio de Janeiro: José Olympio, 1980.

Lins do Rego, José. *O moleque Ricardo.* Rio de Janeiro: José Olympio, 1935.

Logan, William N. *Soil Map of Mississippi.* Jackson: Mississippi Geological Survey, 1927. n.p.

López, Silvia L. "National Culture, Globalization, and the Case of Post-War El Salvador." *Comparative Literature Studies* 41.1 (2004): 80-99.

Luce, Dianne C. *Annotations to Faulkner's* As I Lay Dying. New York: Garland, 1990.

Lynch, Kevin. *The Image of the City.* Cambridge: MIT P and Harvard UP, 1960.

Lynch, Kevin. *What Time is This Place?* Cambridge: MIT P, 1972.

Madden, Lori. "The Canudos War in History." *Luso-Brazilian Review* 30.2 (1993): 6-22.

Maguire, Robert A. *Exploring Gogol.* Stanford: Stanford UP, 1994.

Malik, Madhu. "*Vertep* and the Sacred/Profane Dichotomy in Gogol's *Dikanka* Stories." *Slavic and East European Journal* 34.3 (1990): 332-47.

Mallory, William E., and Paul Simpson-Housley, eds. *Geography and Literature.* Syracuse: Syracuse UP, 1987.

Manguel, Angelo, and Gianni Guadalupi. *The Dictionary of Imaginary Places.* New York: Harcourt Brace, 2000.

Margenot, John B. "Cartography in the Fiction of Juan Benet." *Letras Peninsulares* 1 (1988): 331-43.

Mariátegui, José Carlos. *Siete Ensayos de Interpretación de la Realidad Peruana.* Lima: Editorial Amauta 1989.

Mariz, Celso. *Evolução econômica da Paraíba.* João Pessoa: União, 1939.

Marotti, Giorgio. "José Lins do Rego and the Long Season of Memory." *Black Characters in the Brazilian Novel.* By Giorgio Marotti. Los Angeles: Center for Afro-American Studies, 1987. 267-324.

Marshall, J., and Glyndwr Williams. *The Great Map of Mankind.* Cambridge: Harvard UP, 1982.

Masing-Delic, Irene. "Philosophy, Myth, and Art in Turgenev's *Notes of a Hunter.*" *The Russian Review* 50.4 (1991): 437-50.

Massey, Doreen. "A Global Sense of Place." *Space, Place, and Gender.* Ed. Doreen Massey. Minneapolis: U of Minnesota P, 1994. 146-56.

Massey, Doreen. *Space, Place, and Gender.* Minneapolis: U of Minnesota P, 1994.

Matthews, John T. "Recalling the West Indies: From Yoknapatawpha to Haiti and Back." *American Literary History* 16.2 (2004): 238-62.

McClennen, Sophia. *The Dialectics of Exile: Nation, Time, Language, and Space in Hispanic Literatures.* West Lafayette: Purdue UP, 2004.

McClintock, A. "Family Feuds: Gender, Nationalism and the Family." *Feminist Review* 44 (1993): 61-80.

McCullough, Kate. *Regions of Identity: The Construction of America in Women's Fiction, 1885-1914.* Stanford: Stanford UP, 1999.

McHale, Brian. *Postmodernist Fiction.* New York: Methuen, 1987.

McHaney, Thomas L. "The Falkners and the Origin of Yoknapatawpha County: Some Corrections." *Mississippi Quarterly* 25 (1972): 249-64.

Meier, Albert. "Seekranke Betrachtungen auf der Königin der Inseln. J.W. von Goethe's Sizilienerfahrung im Zusammenhang der *Italienischen Reise.*" *Germanisch Romanische Monatsschrift* 39.2 (1989): 180-95.

Melville, Herman. *Moby Dick.* New York: Library of America, 1983.

Meyrowitz, Joshua. *No Sense of Place: The Impact of Electronic Media on Social Behavior.* New York: Oxford UP, 1985.

Michaelsen, Scott, and David E. Johnson, eds. *Border Theory: The Limits of Cultural Politic*. Minneapolis: U of Minnesota P, 1997.

Miller, J. Hillis. *Topographies*. Stanford: Stanford UP, 1995.

Miller, Jim Wayne. "Anytime the Ground is Uneven: The Outlook for Regional Studies." *Geography and Literature*. Ed. William E. Mallory and Paul Simpson-Housley. Syracuse: Syracuse UP, 1987. 1-23.

Millgate, Michael. "Unreal Estate: Reflections on Wessex and Yoknapatawpha." *The Literature of Region and Nation*. Ed. R. Draper. London: MacMillan, 1989. 61-80.

Milum, Richard A. "Continuity and Change: The Horse, the Automobile, and the Airplane in Faulkner's Fiction." *Faulkner: The Unappeased Imagination. A Collection of Critical Essays*. Ed. Glenn O. Carey. Troy: Whitston, 1980. 157-74.

Mitchell, W.J.T., ed. *Landscape and Power.* Chicago: U of Chicago P, 1994.

Monmonier, Mark. *Drawing the Line*. New York: Henry Holt, 1995.

Muehrcke, Phillip C., and Juliana O. Muehrcke. "Maps in Literature." *The Geographical Review* 44.3 (1974): 317-38.

Murakami, Haruki. *Hard-Boiled Wonderland and the End of the World*. Trans. Alfred Birnbaum. New York: Kodansha, 1991.

Nelson, Esther W. "Narrative Perspective in *Volverás a Región*." *The American Hispanist* 4.36 (1979): 3-6.

Nietzsche, Friedrich. *Zur Genealogie der Moral*. 1887. Stuttgart: Reclam, 1988.

Nunes, Mark. "Virtual Topographies: Smooth and Striated Cyberspace." *Cyberspace Textuality*. Ed. Marie-Laure Ryan. Bloomington: Indiana UP, 1998. 61-77.

Oberhelman, Harley. "William Faulkner and Gabriel García Márquez: Two Nobel Laureates." *Critical Essays on Gabriel García Márquez*. Ed. George R. McMurray. Boston: Hall, 1987. 67-79.

Oliveira, Franklin de. *Euclydes: A espada e a letra*. Rio de Janeiro: Paz e Terra, 1983.

Pacheco, João. *O mundo que José Lins do Rego fingiu*. Rio de Janeiro: São José, 1958.

Padrón, Ricardo. *The Spacious Word: Cartography, Literature and Empire in Early Modern Spain*. Chicago: U of Chicago P, 2004.

Papastergiardix, Nikos. *The Turbulence of Migration*. Cambridge: Polity P, 2000.

Parker, Alice. "The Mauve Horizon of Nicole Brossard." *Québec Studies* 10 (1990): 107-19.

Parkes, D., and Nigel Thrift. *Times, Spaces, and Places*. New York: John Wiley, 1980.

Parmenter, Barbara McKean. *Giving Voices to Stones*. Austin: U of Texas P, 1991.

Pate, Willard. "Pilgrimage to Yoknapatawpha." *The Furman Magazine* 17.4 (1969): 6-13.

Pendergrast, Garrett Elliott. *A Physical and Topographical Sketch of the Mississippi Territory, Lower Louisiana, and a Part of West Florida*. Philadelphia: Gazette of the United States, 1803.

Perlina, Nina. "Travels in the Land of Cokaigne, Sluggard's Land, and Dikanka." *The Supernatural in Slavic and Baltic Literature: Essays in Honor of Victor Terras*. Ed. Amy Mandelher and Roberta Reeder. Columbus: Slavic Publishers, 1988. 57-71.

Peterkin, Julia. *Scarlet Sister Mary.* New York: Grosset & Dunlap, 1928.

Pike, Burton. *The Image of the City in Modern Literature*. Princeton: Princeton UP, 1981.

Piper, Karen. *Cartographic Fictions*. New Brunswick: Rutgers UP, 2002.

Piretto, Gian Piero. "Uomini e natura nelle *Memorie di un cacciatore* di I.S. Turgenev e in *Vita dei campi* di G. Verga." *Mondo slavo e cultura italiana.* Rome: Il Veltro, 1983. 299-313.

Pitavy, François L. "Idiocy and Idealism: Reflection on the Faulknerian Idiot." *Faulkner and Idealism: Perspectives from Paris*. Ed. Michel Gresset and Patrick H. Samway. Jackson: UP of Mississippi, 1983. 97-111.

Pocock, Douglas C.D., ed. *Humanistic Geography and Literature: Essays on the Experience of Place*. Totowa: Barnes & Noble, 1981.

Pope, Randolph D. "Benet, Faulkner, and Bergson's Memory." *Critical Approaches to the Writings of Juan Benet*. Ed. Roberto C. Manteiga, David K. Hertzberger, and Malcolm A. Compitello. Hanover: UP of New England, 1984. 111-19.

Porter, Carolyn. "Symbolic Fathers and Dead Mothers." *Faulkner and Psychology*. Ed. Donald M. Kartiganer and Ann J. Abadie. Jackson: U of Mississippi P, 1994. 78-122.

Pratt, Mary Louise. *Imperial Eyes: Travel Writing and Transculturation*. New York: Routledge, 1992.

Proença, M. Cavalcanti. "O Monstruoso Anfiteatro." *Estudos Literários*. By M. Cavalcanti Proença. Rio de Janeiro: José Olympio, 1971. 251-67.

Radcliffe, Sarah, and Sallie Westwood. *Remaking the Nation*. New York: Routledge, 1996.

Rama, Angel. *La transculturación narrativa en América latina*. Mexico: Siglo Veintiuno, 1982.

Rapin, Ronald. "Cloisters and Cloisterers in Juan Benet's *Volverás a Región*: The Amorphous Nightmare." *Neophilologus* 73.4 (1989): 541-47.

Ratzel, Friedrich. *Anthropogeographie*. Stuttgart: Engelhorn, 1921.

Relph, E. *Place and Placelessness*. London: Pion, 1976.

Richardson, Daniel C. "Towards Faulkner's Presence in Brazil: Race, History, and Place in Faulkner and Amado." *South Atlantic Review* 65.4 (2000): 13-27.

Rimstead, Roxanne. "*Klee Wyck*: Redefining Region through Marginal Realities." *Canadian Literature* 130 (1991): 29-47.

Risso, Mercedes Sanfelice. "*Fogo morto*: Uma narrativa cênica." *Revista de Letras* 23 (1983): 23-37.

Rivero, José. "Juan Benet: Proyecto y cartografía." *Cuadernos Hispanoamericanos* 651-52 (2004): 147-53.

"Rom." *Rom-Paris-London. Erfahrung und Selbsterfahrung deutscher Schriftsteller und Künstler in den Fremden Metropolen.* Ed. Conrad Wiedemann. Stuttgart: Metzler, 1988. 197-344.

Rosa, Julio M. de la. "Encuentro con Région." *Cuadernos Hispanoamericanos* 369 (1981): 587-92.

Rothe, Wolfgang. *Der politische Goethe.* Göttingen: Vandenhoeck & Ruprecht, 1998.

Rouse, Roger. "Mexican Migration and the Social Space of Postmodernism." *Between Two Words: Mexican Immigrants in the United States.* Ed. David G. Gutiérrez. Wilmington: Scholarly Resources, 1996. 247-63.

Rubin, Louis D., ed. *I'll Take My Stand.* 1930. Baton Rouge: Louisiana State UP, 1977.

Rudnick, Lois. "Re-Naming the Land." *The Desert is No Lady.* Ed. Vera Norwood and Janice Monk. New Haven: Yale UP, 1987. 10-26.

Rupert, James. "Mary Austin's Landscape Line in Native American Literature." *Southwest Review* 68 (1983): 376-90.

Rushdie, Salman. *Midnight's Children.* New York: Knopf, 1981.

Ryden, Kent C. *Mapping the Invisible Landscape: Folklore, Writing, and the Sense of Place.* Iowa City: U of Iowa P, 1993.

Sack, Robert D. *Conceptions of Space in Social Thought: A Geographic Perspective.* Minneapolis: U of Minnesota P, 1980.

Sack, Robert D. *Homo Geographicus.* Baltimore: Johns Hopkins UP, 1997.

Saïd, Edward. *Culture and Imperialism.* New York: Random House, 1993.

Saïd, Edward. "Invention, Memory, and Place." *Landscape and Power.* Ed. W.J.T. Mitchell. Chicago: U of Chicago P, 2002.

Salvadon, Marjorie. *Shifting Boundaries: The Interplay of Place and Displacement in Selected Contemporary Francophone Narrative.* Ph.D. diss. Providence: Brown U, 1996.

Sánchez Ferlosio, Rafael. *El Jarama.* Barcelona: Destino, 1961.

Sánchez Ferlosio, Rafael. *The River.* Trans. Margaret Jull Costa. Sawtry: Dedalus, 2004.

Saunders, D.B. "Contemporary Critics of Gogol's *Vechera* and the Debate about Russian *narodnost'* (1831-1832)." *Harvard Ukrainian Studies* 5.1 (1981): 66-82.

Schama, Simon. *Landscape and Memory.* New York: Knopf, 1995.

Schlösser, Hermann. *Reiseformen des Geschriebenen.* Wien: Böhlau, 1987.

Schmitt, Carl. *Land und Meer. Eine Weltgeschichtliche Betrachtung.* Stuttgart: Klett-Cotta, 1993.

Schudt, Ludwig. *Italienreisen im 17. und 18. Jahrhundert.* Wien: Schroll, 1959.

Schwartz, Roberto. *Misplaced Ideas.* Ed. John Gledson. New York: Verso, 1999.

Schwartz, Stuart B. *Sugar Plantations in the Formation of Brazilian Society: Bahaia, 1550-1835.* Cambridge: Cambridge UP, 1985.

Scott, Mack Harris. *"All Our Yesterdays": The Vital Statistics of Yoknapatawpha County.* M.A. thesis. Nashville: Vanderbilt U, 1963.

Shkandrij, Miroslav. *Russia and Ukraine.* Montréal: McGill-Queen's UP, 2001.

Sitwell, O.F.G., and Olenka S.E. Bilash, "Analysing the Cultural Landscape as a Means of Probing the Non-Material Dimension of Reality." *The Canadian Geographer / Le Géographe Canadien* 30.2 (1986): 132-45.

Smith, Bradley. "The Faulkner Country." *'48: The Magazine of the Year* 2.5 (1948): 83-90.

Smith, Jonathan M. "Writing the Aesthetic Experience." *Writing Worlds: Discourse, Text and Metaphor in the Representation of Landscape.* Ed. Trevor J. Barnes and James S. Duncan. New York: Routledge, 1992.

Smyrniv, Walter. "A Gallery of Idealists and Realists." *Critical Essays on Ivan Turgenev.* Ed. David A. Lowe. Boston: Hall, 1989. 73-78.

Soares, Mariana. *O Ontológico na Obra de José Lins do Rego.* João Pessoa: Fundação Espaço Cultural da Paraíba, 1984.

Soja, Edward. *The Political Organization of Space.* Washington: Commission on College Geography Publications, 1971.

Soja, Edward. *Thirdspace.* London: Blackwell, 1996.

Souza Andrade, Olímpio de. *Euclides e o espírito de renovação.* Rio de Janeiro: Livraria São José, 1967.

Späth, Sibylle. *Rolf Dieter Brinkmann.* Stuttgart: Metzler, 1989.

Spitta, Silvia. "José María Arguedas: Entre Dos Aguas." *Between Two Waters.* By Silvia Spitta. Houston: Rice UP, 1995. 139-76.

Standley, Arline R. "Here and There: Now and Then." *Luso-Brazilian Review* 23.1 (1986): 61-75.

Stebelsky, Ihor. "National Identity of Ukraine." *Geography and National Identity.* Ed. David Hooson. Oxford: Blackwell, 1994. 235-38.

Stineman, Ester Lanigan. *Mary Austin: Song of a Maverick.* New Haven: Yale UP, 1989.

Sundquist, Eric J. *Faulkner: The House Divided.* Baltimore: Johns Hopkins UP, 1983.

Taylor, Kit Sims. *Sugar and the Underdevelopment of Northeastern Brazil, 1500-1970.* Gainesville: U of Florida P, 1970.

Taylor, Walter. *Faulkner's Search for a South.* Urbana: U of Illinois P, 1983.

Teague, David W. *The Southwest in American Literature and Art.* Tucson: U of Arizona P, 1997.

Terán, M. de, and L. Sole Sabaris. *Geografía regional de España.* Barcelona: Ariel, 1978.

Terts, Abram [Andrei Sinyavsky]. *V teni Gogolia* (In Gogol's Shadow). London: Collins, 1975.

Thomas, Gareth. *The Novel of the Spanish Civil War.* Cambridge: Cambridge UP, 1990.

Thompson, Ewa M. *Imperial Knowledge: Russian Literature and Colonialism.* Westport: Greenwood P, 2000.

Thrift, Nigel. "Literature, the Production of Culture and the Politics of Place." *Antipode* 15.1 (1983): 12-24.

Tobler, Waldo R. "Geographic Area and Map Projections." *Geographical Review* 53 (1963): 59-78.

Toller Gomes, Heloísa. *O poder do rural na ficção.* São Paulo: Ática, 1981.

Tötösy de Zepetnek, Steven. "From Comparative Literature Today Toward Comparative Cultural Studies." *Comparative Literature and Comparative Cultural Studies.* Ed. Steven Tötösy de Zepetnek. West Lafayette: Purdue UP, 2003. 235-67.

Troyat, Henri. *Turgenev.* Trans. Nancy Amphoux. London: Allison & Busby, 1991.

Tuan, Yi-fu. *Cosmos and Hearth.* Minneapolis: U of Minnesota P, 1996.

Tuan, Yi-fu. "Desert and Ice: Ambivalent Aesthetics." *Landscape, Natural Beauty and the Arts.* Ed. Salim Kemal and Ivan Gaskell. Cambridge: Cambridge UP, 1993. 139-57.

Tuan, Yi-fu. "Language and the Making of Place: A Narrative-Descriptive Approach." *Annals of the Association of American Geographers* 81.4 (1992): 384-96.

Tuan, Yi-fu. *Man and Nature.* Washington: Association of American Geographers, 1971.

Tuan, Yi-fu. *Space and Place: The Perspective of Experience.* Minneapolis: U of Minnesota P, 1977.

Turgenev, Ivan. "Letter to Aleksandr Gertsen of 23 October/4 November 1862." *Pisma' v 'vosemnadtsati tomakh.* Vol. 6. Moscow: Izd-vo 'nauka, 1982. 123-24.

Turgenev, Ivan. *Sketches from a Hunter's Album.* Trans. Richard Freeborn. London: Penguin, 1990.

Turgenev, Ivan. *Zapiski oxotnika.* 1852. Moscow: Akademia Nauk, 1963.

Twain, Mark. *Tom Sawyer Abroad.* New York: Oxford UP, 1996.

United Nations High Commission on Refugees. "Refugees by the Numbers." (2006): <http://www.unhcr.org/cgibin/texis/vtx/basics/opendoc.htm?tbl=BASICS&id=3b028097c>.

Vargas Llosa, Mario. *Entre sapos y halcones.* Madrid: Ediciones Cultura Hispánica del Centro Iberoamericano de Cooperacion, 1978.

Vargas Llosa, Mario. *La utopia arcáica: Jose Maria Arguedas y las ficciones del indigenismo.* Mexico: Fondo de Cultura Económica, 1996.

Vila, Marco Aurelio. *Lo geográfico en* Doña Barbara. Caracas: Ministerio de Relaciones Exteriores, 1986.

Virilio, Paul. "The State of Emergency." *The Paul Virilio Reader.* Ed. James der Derian. Malden: Blackwell, 1999. 46-57.

Wagley, Charles. *Race and Class in Rural Brazil.* Paris: UNESCO, 1952.

Walters, Wendy. *Landscapes of Identity: Figures of Home and Elsewhere Among Writers of the African Diaspora.* Ph.D. diss. San Diego: U of California San Diego, 1996.

Warren, Marsha. "Time, Space, and Semiotic Discourse in the Feminization/Disintegration of Quentin Compson." *The Faulkner Journal* 4.1-2 (1988-89): 99-111.

Wegner, Phillip E. *Imaginary Communities: Utopia, the Nation, and the Spatial Histories of Modernity*. Berkeley: U of California P, 2002.

Werner, Hans-Georg. "Goethes Reise durch Italien als soziale Erkundung." *Goethe Jahrbuch* 105 (1988): 27-41.

White, Hayden. *Tropics of Discourse*. Baltimore: Johns Hopkins, 1985.

Williams, Raymond. *The Country and the City*. New York: Oxford UP, 1973.

Wolff, Maria Tai. "'Estas páginas sem brilhos': O Texto-Sertão de Euclydes da Cunha." *Revista Iberoamericana* 50.126 (1984): 47-61.

Wyatt, David. "Mary Austin: Nature and Nurturance." *The Fall into Eden*. By David Wyatt. New York: Cambridge UP, 1986. 67-95.

Zamora, Lois Parkinson. *The Imagination of History in Recent Fiction of the Americas*. Cambridge: Cambridge UP, 1997.

Zerubavel, Eviatar. *Terra Cognita*. New Brunswick: Rutgers UP, 1997.

Index